Niek Yoan (Ed.)

Samsung Galaxy Ace 2

Niek Yoan (Ed.)

Samsung Galaxy Ace 2

Smartphone, Android (operating system), Samsung

Miss Press

Imprint

Publisher:
Miss Press is a trademark of
International Book Market Service Ltd., 17 Rue Meldrum, Beau Bassin, 1713-01 Mauritius
Email: info@bookmarketservice.com
Website: www.bookmarketservice.com

Published in 2012

Printed in: U.S.A., U.K., Germany. This book was not produced in Mauritius.

ISBN: 978-620-1-49947-8

Contents

Articles

References

Samsung_Galaxy_Ace_2

Galaxy Ace 2
GT-I8160

Manufacturer	Samsung Electronics
Series	Galaxy
Compatible networks	**GSM** 850/900/1800/1900 **HSPA** 14.4(DL)/5.76(UL) Mbps 900/2100 Mhz (Canada **HSPA** 850/1900)
First released	**UK** May 22, 2012
Predecessor	GT-S7500
Type	Smartphone
Form factor	Slate (can be converted)
Dimensions	112.4 mm (**unknown operator: u'strong'** in) H 59.9 mm (**unknown operator: u'strong'** in) W 11.5 mm (**unknown operator: u'strong'** in) D
Weight	113 g (**unknown operator: u'strong'** oz)
Operating system	Android 2.3.6 Gingerbread
CPU	Dual-core 800 MHz , NovaThor U8500
GPU	Mali-400MP
Memory	768 MB RAM(accessible: 555 MB)
Storage	4 GB
Removable storage	microSD (supports up to 32 GB)
Battery	1,500 mAh, 5.0 Wh, 3.7 V, internal rechargeable Li-ion, user replaceable
Data inputs	Multi-touch, capacitive touchscreen Accelerometer A-GPS GLONASS Digital compass Proximity sensor Push buttons Capacitive touch-sensitive buttons
Display	PLS TFT LCD, 3.8 in (**unknown operator: u'strong'** mm) diagonal. 480×800 px WVGA with Gorilla Glass 16M colors
Rear camera	5.0 megapixel 2560×1920 max, 16x digital zoom, autofocus, LED flash, HD video recording 1280×720 px MPEG4 at 30 fps max.
Compatible media formats	**Audio** MP3, AAC, AAC+, eAAC+ **Video** MP4, H.264, H.263
Ringtones & notifications	Vibration, MP3, WAV
Connectivity	3.5 mm jack Bluetooth v3.0 with A2DP DLNA Stereo FM radio with RDS Micro-USB 2.0 Wi-Fi 802.11 b/g/n

Other	Gorilla glass display, Swype keyboard
SAR	0.84 W/kg

The **Samsung Galaxy Ace 2**, is a smartphone manufactured by Samsung that runs the open source Android operating system. Announced and released by Samsung in February 2012, the Galaxy Ace 2 is the successor to the Galaxy Ace Plus,.[1]

The Galaxy Ace 2 is one of the budget, dual core Android-based smartphones planned by Samsung. Being an upper mid-range smartphone, the Ace 2 has a hardware between that of the Galaxy Ace Plus and the Galaxy S Advance; it features a Dual-core 800 MHz processor on NovaThor U8500 chipset with the Mali-400MP as GPU.

It is currently available in black and white.

In May 2012 it went on sale in the UK with four carriers including Orange and 3.[2]

Features

The Galaxy Ace 2 is a 3.5G smartphone, offers quad-band GSM and announced with two-band HSDPA (900/2100)MHz at 14.4(DL)/5.76(UL) Mbit/s. The display is a 3.8 inch PLS TFT LCD capacitive touchscreen of WVGA (480x800) resolution. There is also a 5-megapixel autofocus camera with LED flash, capable of recording videos at QVGA (320x240), VGA (640x480) and HD (1280x720) pixels resolution. The Ace 2 comes with a 1500mAh Li-Ion battery.[3] The Ace 2 runs Android 2.3.6 Gingerbread, Samsung's proprietary TouchWiz 4.0 user interface.

Ace has heavy social network integration abilities and multimedia features. It also is preloaded with Google Trademark Mobile Apps

References

[1] "Play and Share Faster and Smarter with the GALAXY Ace Plus" (http://www.samsungmobilepress.com/2012/01/03/Play-and-Share-Faster-and-Smarter-with-the-GALAXY-Ace-Plus). Samsung View. 2012-01-03. . Retrieved 2012-01-03.

[2] "SAMSUNG GALAXY ACE 2" (http://www.samsung.com/uk/consumer/mobile-devices/smartphones/android/GT-I8160OKABTU). Samsung View. 2012-01-05. . Retrieved 2012-01-05.

[3] "Samsung Galaxy Ace 2 S8160 - Full phone specifications" (http://www.gsmarena.com/samsung_galaxy_ace_2_i8160-4559.php). Gsmarena.com. . Retrieved 2012-02-13.

Smartphone

A **smartphone** is a mobile phone built on a mobile computing platform, with more advanced computing ability and connectivity than a feature phone.[1][2][3] The first smartphones mainly combined the functions of a personal digital assistant (PDA) and a mobile phone or camera phone. Today's models also serve to combine the functions of portable media players, low-end compact digital cameras, pocket video cameras, and GPS navigation units.

Modern smartphones typically also include high-resolution touchscreens, web browsers that can access and properly display standard web pages rather than just mobile-optimized sites, and high-speed data access via Wi-Fi and mobile broadband. The most common mobile operating systems (OS) used by modern smartphones include Google's Android, Apple's iOS, Nokia's Symbian, RIM's BlackBerry OS, Samsung's Bada, Microsoft's Windows Phone, Hewlett-Packard's webOS, and embedded Linux distributions such as Maemo and MeeGo. Such operating systems can be installed on many different phone models, and typically each device can receive multiple OS software updates over its lifetime.

The distinction between smartphones and feature phones can be vague and there is no official definition for what constitutes the difference between them. One of the most significant differences is that the advanced application programming interfaces (APIs) on smartphones for running third-party applications[4] can allow those applications to have better integration with the phone's OS and hardware than is typical with feature phones. In comparison, feature phones more commonly run on proprietary firmware, with third-party software support through platforms such as Java ME or BREW.[1] An additional complication in distinguishing between smartphones and feature phones is that over time the capabilities of new models of feature phones can increase to exceed those of phones that had been promoted as smartphones in the past.

Some manufacturers use the term "superphone" for their high end phones with unusually large screens and other expensive features.[5][6] Other commentators prefer "phablet" in recognition of their convergence with low-end tablet computers.[7][8]

About 18% of total world population possess some sort of smartphone by 2012, compared to 12% in 2010 and 8% in 2008.

History

Origin of the term

Although devices combining telephony and computing were conceptualized as early as 1973 and were offered for sale beginning in 1994, the term "smartphone" did not appear until 1997, when Ericsson described its GS 88 "Penelope" concept as a "Smart Phone".[9][10][11][12][13]

Early years

As early as 1973, 8 years prior to the availability of the modern cellular telephone and 21 years prior to the sale of the first smartphone, Theodore George "Ted" Paraskevakos introduced the concepts of combining intelligence, data processing and visual display screens with telephones. In a foreshadowing of futuristic capabilities that would eventually become commonplace, Paraskevakos' early patents covered the concepts of banking and paying utility bills via telephone.[9][14]

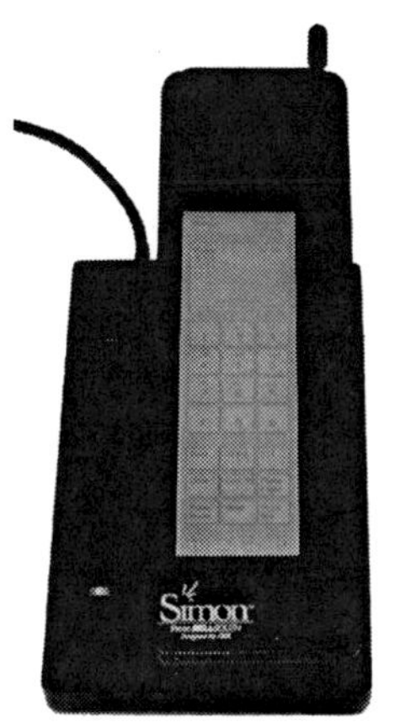

IBM Simon and charging base (introduced August 16, 1994)

The first cellular phone to include telephony and PDA features in one device was an International Business Machines Inc. (IBM) prototype developed in 1992 that was code named "Angler". It was shown as a concept product that year at the COMDEX computer industry trade show. A refined version of the product was announced the following year by BellSouth Cellular Crop. as the Simon Personal Communicator and marketed to consumers beginning on August 16, 1994. The Simon was the first device available to consumers that can be properly referred to as a "smartphone", although the term had not be coined at the time the Simon went on sale.[10][15] In addition to its ability to make and receive cellular phone calls, Simon was also able to send and receive facsimiles, e-mails and pages through its touch screen display. Simon included many applications including an address book, calendar, appointment scheduler, calculator, world time clock, games, electronic note pad, handwritten annotations and standard and predictive touchscreen keyboards.

The Nokia 9000, released in 1996 and part of the Nokia Communicator model line, was the first of Nokia's smartphones. This distinctive palmtop computer-style smartphone combined a costly Hewlett-Packard personal digital assistant (PDA) with Nokia's best-selling phone of that time. In early prototypes, the two devices were fixed together via a hinge in what came to be described as a clamshell design. These Nokia devices have a feature phone display, keyboard and user interface on the inside top surface of the opened phone with a physical QWERTY keyboard, high-resolution display of at least 640×200 pixels and PDA user interface under inside the bottom surface of the opened phone. The Nokia Communicators featured email communication and text-based web browsing via their GEOS V3.0 operating system.

In the late 1990s the vast majority of mobile phones had only basic phone features and many people who needed functionality beyond that also carried PDA and/or pager type devices running early versions of operating systems such as Palm OS, BlackBerry OS or Windows CE/Pocket PC.[1] Later versions of these systems started integrating cell phone capabilities with their PDA and messaging features and support of third-party applications. Today, high-end devices running these systems are often branded smartphones.

In early 2001, Palm, Inc. introduced the Kyocera 6035, the first smartphone to be deployed in widespread use in the United States. This device combined the features of a personal digital assistant (PDA) with a wireless phone that operated on the Verizon Wireless network. For example, a user could select a name from the PDA contact list, and the device would dial that contact's phone number. The device also supported limited web browsing.[16] The device received a very positive reception from technology publications, but the product line never became widespread outside North America.[17]

Operating systems

Symbian

In 2000, the touchscreen Ericsson R380 Smartphone was released.[18] It was the first device to use an open operating system, the Symbian OS.[19] It was the first device marketed as a 'smartphone'.[20] It combined the functions of a mobile phone and a personal digital assistant (PDA).[21] In December 1999 the magazine Popular Science called the Ericsson R380 Smartphone one of the most important advances in science and technology.[22] It was groundbreaking in being as small and light as a normal mobile phone.[23] In 2002 it was followed up by P800.[24]

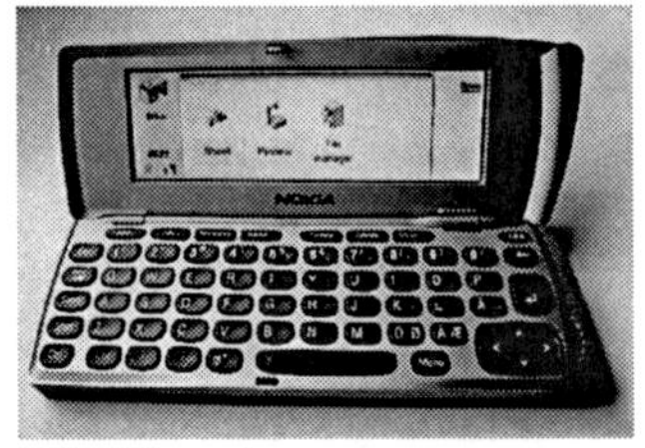

The Nokia 9210 Communicator (introduced November 21, 2000)

Also in 2000, the Nokia 9210 communicator, the first color screen model from the Nokia Communicator line, also had Symbian. The later 9500 was Nokia's first camera phone and first Wi-Fi phone. The 9300 was smaller, and the latest E90 Communicator includes GPS. The Nokia Communicator series was the most costly phone model sold by a major brand for almost the full life of the series, costing 20% and sometimes 40% more than the next most expensive smartphone by any major producer.

In 2007, Nokia launched the Nokia N95 which integrated various multimedia features into a consumer-oriented smartphone: GPS, a 5 megapixel camera with autofocus and LED flash, 3G and Wi-Fi connectivity and TV-out. In the next few years these features would become standard on high-end smartphones. The Nokia 6110 Navigator is a Symbian based dedicated GPS phone introduced in June 2007.

In 2010, Nokia released the Nokia N8 smartphone with a stylus-free capacitive touchscreen, the first device to use the new Symbian^3 OS.[25] Its megapixel camera able to record HD video in 720p was described by Mobile Burn as the best camera in a phone,[26] and Mobile Choice described its satellite navigation as the best on any phone.[27] It also featured a front-facing VGA camera for videoconferencing.

Symbian was the number one smartphone platform by market share from 1996 until 2011 when it dropped to second place behind Google's Android OS. In February 2011, Nokia announced that it would replace Symbian with Windows Phone as the operating system on all of its future smartphones.[28] This transition was completed in October 2011, when Nokia announced its first line of Windows Phone 7.5 smartphones, Nokia Lumia 710 and Nokia Lumia 800.[29]

Nokia is committed to support its Symbian based smartphones until 2016, continuing to release further OS improvements, like Nokia Belle and Nokia Belle FP1, and new devices, like the award winning Nokia 808 PureView.

Windows

The HTC Touch Pro2 smartphone (introduced May 2009)

In 2001, Microsoft announced its Windows CE Pocket PC OS would be offered as "Microsoft Windows Powered Smartphone 2002."[30] Microsoft originally defined its Windows Smartphone products as lacking a touchscreen and offering a lower screen resolution compared to its sibling Pocket PC devices.

Palm OS

In early 2002, Handspring released the Palm OS Treo smartphone, utilizing a full keyboard that combined wireless web browsing, email, calendar, and contact organizer with mobile third-party applications that could be downloaded or synced with a computer.[31]

BlackBerry

In 2002, RIM released their first BlackBerry devices with integrated phone functionality and shifted the positioning of their products from 2-way pagers to email-capable mobile phones. The BlackBerry line evolved into the first smartphone optimized for wireless email use and had achieved a total customer base of about 32 million subscribers by December 2009.[32]

iOS

First generation Apple iPhone (introduced June 2007)

In 2007, Apple Inc. introduced its costly iPhone, one of the first mobile phones to use a multi-touch interface. The iPhone was notable for its use of a large touchscreen for direct finger input as its main means of interaction, instead of a stylus, keyboard, and/or keypad as typical for smartphones at the time. *Ars Technica* described its Web browser as "far superior" to anything offered by that of its competitors.[33] Initially lacking the capability to install native applications beyond the ones built-in to its OS, at WWDC in June 2007 Apple announced that the iPhone would support third-party "web 2.0 applications" running in its web browser that share the look and feel of the iPhone interface.[34] As a result of the iPhone's initial inability to install third-party native applications, some reviewers did not regard it as a smartphone "by conventional terms."[35] A process called jailbreaking emerged quickly to provide unofficial third-party native applications. The different functions of the iPhone (including a GPS unit, kitchen timer, radio, map book, calendar, notepad, and many others) allowed consumers to replace all of these items.[36]

In July 2008, Apple introduced its second generation iPhone with a much lower list price and 3G support. Simultaneously, also created the App Store, adding the capability for any iPhone or iPod Touch to install, directly and officially, additional native applications (both free and paid) over a Wi-Fi or cellular network, without the more typical process at the time of requiring a PC for installation. Applications could additionally be browsed through and downloaded directly via the iTunes software client on Macintosh and Windows PCs, rather than by searching through multiple sites across the Internet. Featuring over 500 applications at launch,[37] Apple's App Store was immediately very popular,[38] quickly growing to become a huge success.[39][40]

In June 2010, Apple introduced iOS 4, which included APIs to allow third-party applications to multitask,[41] and the iPhone 4, which included a 960×640 pixel display with a pixel density of 326 pixels per inch (ppi), a 5 megapixel camera with LED flash capable of recording HD video in 720p at 30 frames per second, a front-facing VGA camera for videoconferencing, an 800 MHz processor, and other improvements.[42] In early 2011 the iPhone 4 became available through Verizon Wireless, ending AT&T's exclusivity of the handset in the U.S.,[43][44][45] and allowing the handset's 3G connection to be used as a wireless Wi-Fi hotspot for the first time, to up to 5 other devices.[46] Software updates subsequently added this capability to other iPhones running iOS 4.[47][48]

The iPhone 4S was announced on October 4, 2011, improving upon the iPhone 4 with a dual core A5 processor, an 8 megapixel camera capable of recording 1080p video at 30 frames per second, World phone capability allowing it to work on both GSM & CDMA networks, and the Siri automated voice assistant.[49] On October 10, Apple announced that over one million iPhone 4Ss had been pre-ordered within the first 24 hours of it being on sale, beating the 600,000 device record set by the iPhone 4,[50][51] despite the iPhone 4S failing to impress some critics at the announcement[52][53] due to their expectations of an "iPhone 5" with rumored drastic changes compared to the iPhone 4 such as a new case design and larger screen.[54] Along with the iPhone 4S Apple also released iOS 5 and iCloud, untethering iOS devices from Macintosh or Windows PCs for device activation, backup, and synchronization,[55] along with additional new and improved features.[56]

Android

Android is an open-source platform backed by Google, along with major hardware and software developers (such as Intel, HTC, ARM, Motorola and Samsung, to name a few), that form the Open Handset Alliance.[57] The first phone to use Android was the HTC Dream, branded for distribution by T-Mobile as the G1. The software suite included on the phone consists of integration with Google's proprietary applications, such as Maps, Calendar, and Gmail, and a full HTML web browser. Android supports the execution of native applications and a preemptive multitasking capability (in the form of services). Third-party apps are available via Google Play (released October 2008 as Android Market), including both free and paid apps.

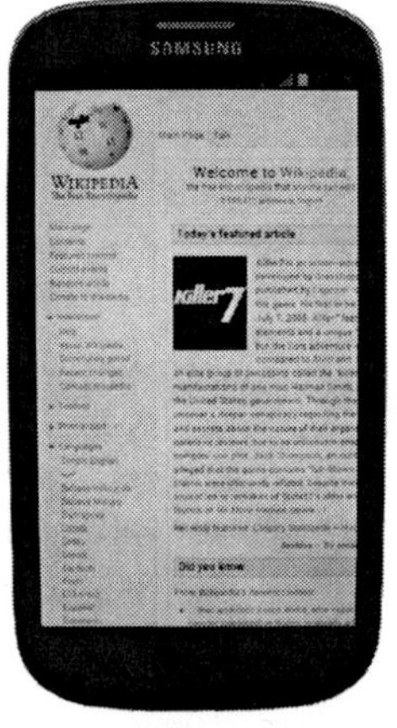

Galaxy S III, the latest smartphone to run this operating system

In January 2010, Google launched the Nexus One smartphone using its Android OS. Although Android has multi-touch abilities, Google initially removed that feature from the Nexus One,[58] but it was added through a firmware update on February 2, 2010.[59]

Concerning the Xperia Play smartphone, an analyst at CCS Insight said in March 2011 that "Console wars are moving to the mobile platform".[60] In the same month, the HTC EVO 3D was announced by HTC Corporation, which can produce 3D effects with no need for special glasses (autostereoscopy).[61] The HTC EVO 3D was officially released on June 24, 2011.[62]

Bada

The Bada operating system for smartphones was announced by Samsung on 10 November 2009.[63][64] The first Bada-based phone was the Samsung Wave S8500, released on June 1, 2010,[65][66] which sold one million handsets in its first 4 weeks on the market.[67]

Samsung shipped 3.5 million phones running Bada in Q1 of 2011.[68] This rose to 4.5 million phones in Q2 of 2011.[69]

Open-source development

The open-source culture has penetrated the smartphone market in several ways. There have been attempts to create open source hardware and software for smartphones.

In February 2010, Nokia made Symbian open source. Thus, most commercial smartphones were based on open-source operating systems. These include those based on Linux, such as Google's Android, Nokia's Maemo, Hewlett-Packard's webOS, and those based on BSD, such as the Darwin-based Apple iOS. Maemo was later merged with Intel's project Moblin to form MeeGo.[70][71]

Malicious software attacks

As smartphone adoption goes up they have increasingly become subject to attacks by malicious software (malware).[72][73]

Frequently this malware is distributed through application stores that have minimal or no review process for their content.[74] In some cases malware has been hidden in pirated versions of legitimate apps, which are then distributed through 3rd party app stores.[75][76] Malware risk also comes from what's known as an "update attack," where a legitimate application is later changed to include a malware component, which users then install when they are notified that the app has been updated. Additionally, the ability to acquire software directly from links on the web results in a distribution vector called "malvertizing," where users are directed to click on links, such as on ads that look legitimate, which then open in the device's web browser and cause malware to be downloaded and installed automatically.[77]

Typical smartphone malware leverages platform vulnerabilities that allow it to gain root access on the device in the background. Using this access the malware installs additional software to target communications, location, or other personal identifying information. A common form of malware on mobile phones is the SMS trojan, which sends premium SMS messages, possibly while unknowingly running in the background of a legitimate application. These premium SMS messages run up charges on the owner's phone bill which cannot be recovered.

In August 2010, Kaspersky Lab reported detection of the first malicious program for smartphones running on Google's Android operating system, named Trojan-SMS.AndroidOS.FakePlayer.a, an SMS trojan which had already infected a number of devices using that OS.[78] Over the spring of 2011 Android malware increased 76%, according to McAfee.[72][79] A report from Juniper Global Threat Center notes that malware on the Android platform increased 400% from 2009 to the summer of 2010, and then saw a 472% increase between July and November 2011.[74] The Juniper report indicates that 55% of Android malware acts as spyware, and 44% are SMS trojans.

While there have been and continue to be potential security flaws in iOS,[80] as of at least August 2011 there were no known malware or spyware apps in Apple's App Store, according to security firm Lookout. There are however commercial spyware applications available, outside the App Store, for jailbroken iOS devices.[77] In June 2011 Symantec's 23-page report "A Window Into Mobile Device Security" characterized (non-jailbroken) devices running iOS as having "full protection" against malware attacks.[81]

Symbian and older versions Windows Mobile have had to contend with a degree of malware in the past, but as legacy systems it is believed that the people who previously targeted them have shifted their focus to Android.[74] There were also a few Palm OS viruses.

The only mobile platform other than Apple's iOS without reports of malware so far is HP's (formerly Palm's) webOS, but this may be explained by its relatively low adoption rate.[79]

The best way to reduce a device's vulnerability to malware attacks is to install the most recent versions of operating systems which include security patches. This can be complicated by long delays[82] in software updates for many devices which have had their software modified with custom "skins," services, or promotional on-deck apps by their manufacturer or mobile carrier.[77] In some cases a device may no longer be receiving updates from its manufacturer or carrier, leaving it vulnerable to exploits that have been patched in an OS version that's more recent than the device's last supported one.

Application stores

The introduction of Apple's App Store for the iPhone and iPod Touch in July 2008 popularized manufacturer-hosted online distribution for third-party applications focused on a single platform. Before this, smartphone application distribution was largely dependent on third-party sources providing applications for multiple platforms, such as GetJar, Handango, Handmark, PocketGear, and others.

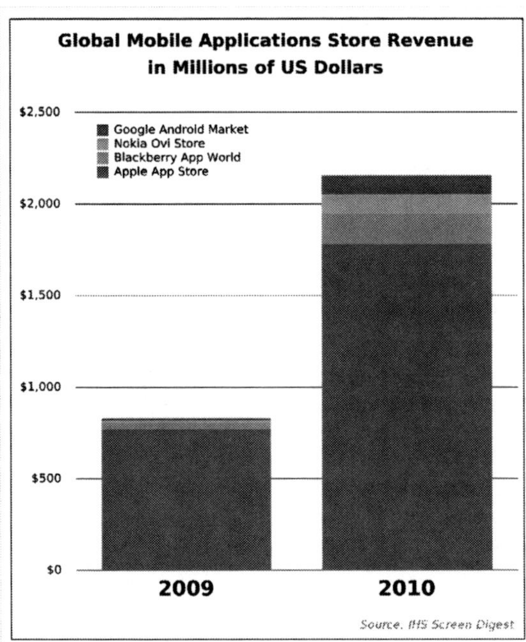

The iPhone's platform is officially restricted to installing apps through the App Store, through "B2B" deployment, and on an "Ad Hoc" basis on up to 100 iPhones.[83] Through jailbreaking it can install apps from other sources. Other platforms may allow application distribution through additional sources outside of their manufacturer-provided app stores, such as third-party app stores and downloads from individual websites.

Following the success of Apple's App Store other smartphone manufacturers quickly launched application stores of their own. Google launched the Android Market in October 2008. MiKandi launched the first adult app market for smartphones in 2009. RIM launched its app store, BlackBerry App World, in April 2009. Nokia launched its Ovi Store in May 2009. Palm launched its Palm App Catalog for webOS in June 2009. Microsoft launched an application store for Windows Mobile called Windows Marketplace for Mobile in October 2009, and then a separate Windows Phone Marketplace for Windows Phone in October 2010. Samsung launched Samsung Apps for its Bada based phones in June 2010. Amazon launched its Amazon Appstore for the Google Android operating system in March 2011.

Store	2009 (millions U.S.)	2010 (millions U.S.)[84]
Apple App Store	$769	$1782
Blackberry App World	$36	$165
Nokia Ovi Store	$13	$105
Google Play	$11	$102
Total	$828	$2155

The relatively high revenue of U.S. $5782 million in 2012 for Apple's App Store compared to competitor's stores[84] can be attributed to a combination of factors. In large part this can be attributed to having the largest number of apps available and the highest download volume of any mobile app store in 2010, but besides that only 28% of the apps in Apple's App Store were free apps, compared to over 57% in the Android Market. Similarly, Nokia's Ovi Store and the BlackBerry App World both had only 26% of their apps available for free, but both generated higher revenues than the Android Market despite having much lower download volumes.[85]

Patent licensing and litigation

The rate of lawsuits, trade complaints, and countersuits and complaints based on patents and designs, in the markets for smartphones and devices based on smartphone OSes such as Android and iOS, increased greatly in 2010. A smartphone war between Samsung and Apple started when Apple claimed that the original Galaxy S Android phone copied its iOS in terms of interface and possibly the hardware of iPhone 3GS.

Features and applications

Display

Screens on smartphones vary largely in both display size and display resolution. The most common screen sizes range from 2 inches to over 4 inches (measured diagonally). Some 5 inch screen devices exist that run on mobile OSes and have the ability to make phone calls, such as the discontinued Dell Streak and the current Samsung Galaxy Note. Ergonomics arguments have been made that increasing screen sizes start to negatively impact usability.

Common resolutions for smartphone screens vary from 240×320 to 720×1280, with many flagship Android phones at 480×800 or 540×960, the iPhone 4/4S at 640×960 and Galaxy Nexus and HTC Rezound at 720×1280.

Popular applications

According to a ComScore report released on May 12, 2011, nearly one in five smartphone users are tapping into check-in services like Foursquare and Gowalla. A total of 16.7 million mobile-phone subscribers used location-based services on their phones in March 2011.[86]

Some smartphones allow watching television[87] using or providing a second screen for media multitasking.[88]

Market share

Smartphones

For several years, the demand for advanced mobile devices boasting powerful processors and graphics processing units, abundant storage (flash memory) for applications and media files, high-resolution screens with multi-touch capability, and open operating systems has outpaced the rest of the mobile phone market.[89]

According to an early 2010 study by ComScore, over 45.5 million people in the United States owned smartphones out of 234 million total subscribers.[90] Despite the large increase in smartphone sales in the last few years, smartphone shipments only made up 20% of total handset shipments as of the first half of 2010.[91]

According to Gartner in their report dated November 2010, total smartphone sales doubled in one year and now smartphones represent 19.3% of total mobile phone sales.[92] Smartphone sales increased in 2010 by 72.1% from the prior year, whereas sales for all mobile phones only increased by 32%.[93][94]

According to an Olswang report in early 2011, the rate of smartphone adoption is accelerating: as of March 2011 22% of UK consumers had a smartphone, with this percentage rising to 31% amongst 24- to 35-year-olds.[95]

In March 2011, Berg Insight reported data that showed global smartphone shipments increased 74% from 2009 to 2010.[96]

A survey of mobile users in the United States by Nielsen in Q3, 2011 reports that smartphone ownership has reached 43% of all U.S. mobile subscribers, with the vast majority of users under the age of 44 owning one. In the 25–34 age range smartphone ownership is reported to be at 62%.[97] NPD Group reports that the share of handset sales that were smartphones in Q3, 2011 reached 59% for consumers 18 and over in the U.S.[98]

In profit share worldwide smartphones now far exceed the share of non-smartphones. According to a November 2011 research note from Canaccord Genuity, Apple Inc. holds 52% of the total mobile industry's operating profits, while only holding 4.2% of the global handset market. HTC and RIM similarly only make smartphones and their worldwide profit shares are at 9% and 7%, respectively. Samsung, in second place after Apple at 29%, makes both smartphones and feature phones and doesn't report a breakdown separating their profits between the two kinds of devices, but it can be intuited that a significant portion of that profit comes from their flagship smartphone devices.[99]

Up to the end of November 2011, camera-equipped smartphones took 27% of photos, a significant increase from 17% in 2010. For many people, smartphones have replaced Point-and-shoot cameras.[100]

Manufacturers

From the launch of their Communicator model in 1996 until 2011 Nokia was dominant in the smartphone market, though has more recently been joined by other competitors in the market. Based on a report by Strategy Analytics, Samsung overtook Nokia in smartphone shipments with an estimated 27.8 million units shipped in Q3 2011[101] (Samsung does not publicly disclose the numbers of their smartphone shipments and sales).

Quantity market shares by Strategy Analytics (note that Gartner instead shows Nokia ahead on Apple sales based on Symbian sales only[102])
(new sales)

Manufacturer	Percent
Apple Q2 2010	13.5%
Apple Q2 2011	18.5%
Samsung Q2 2010	5.0%
Samsung Q2 2011	17.5%
Nokia Q2 2010	38.1%
Nokia Q2 2011	15.2%
Others Q2 2010	43.4%
Others Q2 2011	48.9%

Market share among smartphone manufacturers does not resemble smartphone OS market share numbers due to the differences between the two major smartphone OS sales models: single manufacturer and licensed. Apple's iPhone, Nokia's Symbian, and RIM's BlackBerry smartphones are currently only available from single manufacturers. Google's Android OS and Microsoft's mobile OSes are platforms that are licensed and used by a variety of manufacturers. As a result, manufacturers of smartphones using licensed OSes all split the total market share of that OS between them, while the total share for a single-manufacturer OS is held by that manufacturer alone.

Note that Nokia's Symbian OS was previously available from several manufacturers under a licensed model, then later predominantly only by Nokia itself more like a single manufacturer model.

Samsung smartphones use a diverse portfolio of operating systems, including their own Bada operating system along with Android and Windows Mobile.[103]

Apple surpassed Nokia worldwide by revenue and profit for the first time in Q2 2011 (though not in market share), with Apple's profit share of the total worldwide smartphone market increasing to 66.3% while Nokia reported a loss.[104]

Between Q2 2010 and Q2 2011 Nokia's worldwide Symbian smartphone sales dropped significantly from 38.1% to 15.2%, while Samsung smartphone sales increased significantly worldwide from 5% to 17.5%.[105] As of Q1 2011,

Nokia had already announced plans to switch to Windows Phone.

Smartphone Customer Satisfaction by J.D. Power and Associates	
Manufacturer	**Score**
Apple 2010	810
Apple 2011	838
HTC 2010	727
HTC 2011	801
Industry Average 2010	753
Industry Average 2011	788
Samsung 2010	724
Samsung 2011	777
Motorola 2010	N/A
Motorola 2011	775
RIM 2010	741
RIM 2011	762
LG 2010	N/A
LG 2011	760
HP/Palm 2010	712
HP/Palm 2011	733
Nokia 2010	720
Nokia 2011	721

Rankings are based on a possible top score of 1000

Nokia remained the number one company in the worldwide mobile phone market with sales for Q2 2011 of 88.5 million when including feature phone platforms such as S40, compared with 16.7 million smartphones running Symbian.[106]

According to Nielsen in July 2011, in the United States Apple is the top smartphone manufacturer at 28% of the market, with RIM at 20%. Google Android has 39% of the U.S. market as a whole, but this is split between HTC at 14%, Motorola at 11%, Samsung at 8%, and other remaining manufacturers at 6%. HTC's total share of the U.S. smartphone market actually ties RIM at 20%, since sales of their smartphones running Microsoft's mobile operating systems account for 6% of the total market. Samsung similarly gains 2% of overall U.S. market share due to their sales of Microsoft OS-based smartphones. In contrast to the worldwide market, Nokia's share of U.S. smartphone sales is very small, at only 2%.[107][108] Nielsen's Q3, 2011 survey of mobile users maintains Apple as the top U.S. smartphone maker with a continued 28% of the market, with RIM dropping from 20% to 18%.[97] While Google Android increased in total operating system share from 39% to 43% of the U.S. market, it remains fragmented amongst many different manufacturers. Over the same quarter Microsoft managed a modest gain from 6% to 7% total U.S. smartphone OS share.

Checks with U.S. carriers by technology analyst firm Canaccord Genuity in April and August 2011 found that Apple's iPhone 4 to be consistently the top selling device at AT&T and Verizon. The second most popular spot at AT&T has been maintained by the iPhone 3GS, which was originally released in 2009 (and has never been sold on Verizon). In August 2011 the most popular smartphones on Sprint and T-Mobile in the U.S. were the HTC EVO 3D 4G and HTC Sensation, respectively. The other second most popular smartphones were the Samsung Charge 4G on

Verizon, the Motorola Photon 4G on Sprint, and the HTC myTouch 4G Slide on T-Mobile.[109][110] NPD Group reported that in Q3, 2011 the overall top 5 smartphones by sales across all carriers in the U.S. were, in order: the iPhone 4, iPhone 3GS, HTC EVO 4G, Motorola Droid 3, and Samsung Intensity II.[98]

In Q1 2012, after 14 years in the market, Samsung surpassed Nokia in units sold. Samsung also taking pole position in smartphones with 44.5 million smartphones sold or 30.6 market share, while 35.1 million iPhones sold or 24.1% market share.[111]

Currently the vast majority of smartphones are manufactured in China, Taiwan and Mexico, for companies based in the U.S. (Apple, HP, Motorola), South Korea (LG, Samsung), Canada (RIM), Finland (Nokia), Taiwan (HTC) and the U.K. (Sony Ericsson).

According to global marketing information services firm J.D. Power and Associates smartphones from Apple Inc. have been consistently[112] ranking highest in customer satisfaction,[113][114] with a late 2011 score of 838 out of 1000. Based on the responses to their most recent survey of 6,898 smartphone users, Apple was followed in ranking by HTC (801), Samsung (777), Motorola (775), RIM (762), LG (760), Palm (733), and Nokia (721).[115][116][117][118]

Operating systems

2010 saw the rapid rise of the Google Android operating system from 4% of new deployments in 2009 to 33% at the beginning of 2011 making it share the top position with the since long dominating Symbian OS. The smaller rivals include Blackberry OS, iOS, Samsung's recently introduced Bada, HP's heir of Palm webOS and the Microsoft Windows Phone OS which is now supported by Nokia.

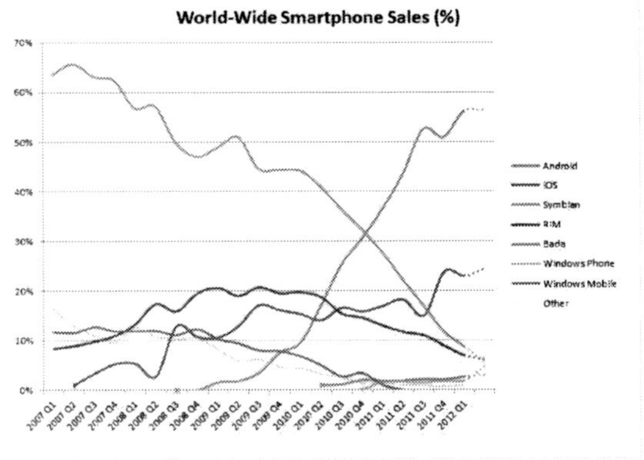

Over late 2009 and 2010 Android's smartphone operating system market share increased very rapidly.[92] In the fourth quarter of 2010, Android surpassed Symbian as the most common operating system in smartphones, with 32.9 million units sold versus 31.0 million. Android-equipped phones sold seven times more than in the prior year.[119] According to Canalys, Google's Android operating system, which is offered to phone makers for free, has raced to the top past operating systems by Nokia, Apple, RIM, and Microsoft. In Q1 2011 Google's Android market share was 35%, increasing significantly from 10% the previous year, while Nokia's Symbian dropped to 26% from 46% over the same time period.[120] In the UK, which currently has one of the highest penetrations of smartphones in the World, Android achieved 50% market share in October 2011.[121]

While carrier subsidies have been responsible for speeding up the growth of smartphone sales, it has been statistically proven that the iPhone's market share is much more heavily dependent on them as compared to competition.[122]

Market Share

Year	Android (Google)	Blackberry (RIM)	IPhone (Apple)	Symbian (Nokia)	Windows Mobile/Phone (Microsoft)	Other
2012-Q1[123]	56.1%	6.9%	22.9%	8.6%	1.9%	3.6%

Historical sales figures

Figures in millions.

Year	Android (Google)	Blackberry (RIM)	IPhone (Apple)	Linux	Palm/WebOS (Palm/HP)	Symbian (Nokia)	Windows Mobile/Phone (Microsoft)	Bada (Samsung)	Other
2007[124]		11.77	3.3	11.76	1.76	77.68	14.7		
2008[124]		23.15	11.42	11.26	2.51	72.93	16.5		
2009[125]	6.8	34.35	24.89	8.13	1.19	80.88	15.03		
2010[126]	67.22	47.45	46.6			111.58	12.38		
2011[127]	219.52	51.54	89.26			93.41	8.77		14.24
2012-Q1[128][127]	81.07	9.94	33.12			12.47	2.71	3.84	1.24

Enterprise share by operating system

In a worldwide study of 2,300 workers at 1,100 businesses by iPass it was reported that Apple's iPhones have displaced RIM's BlackBerry devices in enterprise adoption in 2011.[129] The share for iPhones increased to 45% from 31.1% in 2010, while the Blackberry share dropped to 32.2% from 34.5% in the previous year. Android phones also increased in share, to 21.3% from 11.3% in 2010, exceeding Symbian for the first time, which dropped to 7.4% from 12.4%. Windows Mobile and all other smartphone OSes also dropped in 2011 compared to 2010.[130]

Customer loyalty by operating system

According to a survey of more than 6,000 smartphone users through 2010 by mobile analytics firm Zokem, the top five loyalty scores for smartphone platforms are the iPhone at 73%, followed by Google's Android at 40%, Samsung's Bada at 33%, RIM's BlackBerry at 30%, and Symbian S60 at 23%. Windows Mobile and Palm follow at 10% each. Customer loyalty gauges the likelihood that the user of a smartphone platform whose contract has expired or who has broken or lost their phone will repurchase another one that uses that same platform.[131][132]

Size of Market

The European mobile market, as measured by active subscribers of the top 50 networks is 860 million.[133]

According to a 2012 survey, half of the U.S. mobile consumers own smartphones. Nielsen, figures, as of February, say 49.7% of U.S. mobile owners have smartphones. Nielsen also notes, that at current rate, smartphones could account for 70% of all U.S. mobile devices by next year.[134]

References

[1] "Smartphone" (http://www.phonescoop.com/glossary/term.php?gid=131). *Phone Scoop*. . Retrieved 2011-12-15.

[2] "Feature Phone" (http://www.phonescoop.com/glossary/term.php?gid=310). *Phone Scoop*. . Retrieved 2011-12-15.

[3] Andrew Nusca (20 August 2009). "Smartphone vs. feature phone arms race heats up; which did you buy?" (http://www.zdnet.com/blog/gadgetreviews/smartphone-vs-feature-phone-arms-race-heats-up-which-did-you-buy/6836). ZDNet. . Retrieved 2011-12-15.

[4] "Smartphone definition from PC Magazine Encyclopedia" (http://www.pcmag.com/encyclopedia_term/0,2542,t=Smartphone&i=51537,00.asp). *PC Magazine*. . Retrieved 2011-12-15.

[5] What Makes a Smartphone a Superphone? (http://mashable.com/2010/07/12/superphone/) Maxhable.com

[6] Superphone vs smartphone: what's the difference? (http://www.techradar.com/news/phone-and-communications/mobile-phones/superphone-vs-smartphone-what-s-the-difference--1055564) Techradar.com

[7] Tablet and Phablet Use Rises Steadily (http://www.informationweek.com/news/mobility/smart_phones/240001040) By Eric Zeman, InformationWeek, May 25, 2012

[8] Enter the Phablet, a History of Phone-Tablet Hybrids (http://www.pcmag.com/slideshow/story/294004/enter-the-phablet-a-history-of-phone-tablet-hybrids) by Sascha Segan, PC Magazine, February 13, 2012

[9] U.S. Patent #3,812,296/5-21-1974 (*Apparatus for Generating and Transmitting Digital Information*), U.S. Patent #3,727,003/4-10-1973 (*Decoding and Display Apparatus for Groups of Pulse Trains*), U.S. Patent #3,842,208/10-15-1974 (*Sensor Monitoring Device*)

[10] Sager, Ira (2012-06-29). "Before IPhone and Android Came Simon, the First Smartphone" (http://www.businessweek.com/articles/2012-06-29/before-iphone-and-android-came-simon-the-first-smartphone). *Bloomberg Businessweek*. Bloomberg L.P. ISSN 2162-657X. . Retrieved 2012-06-30. "...Simon wasn't ready for its scheduled release in May 1994. Customers couldn't get one until Aug. 16."

[11] "Ericsson GS88 Preview" (http://pws.prserv.net/Eri_no_moto/GS88_Preview.htm). Eri-no-moto. 2006. . Retrieved 2011-12-15.

[12] "History" (http://www.stockholmsmartphone.org/history/). Stockholm Smartphone. 2010. . Retrieved 2011-12-15.

[13] "Penelope box" (http://www.stockholmsmartphone.org/wp-content/uploads/penelope-box.jpg). . Retrieved 2011-12-15.

[14] "1980–90" (http://www.tekniskamuseet.se/mobilen/engelska/1980_90.shtml). The National Museum of Science and Technology. 2002. . Retrieved 2009-07-29.

[15] Schneidawind, John (1992-11-23). "Poindexter putting finger on PC bugs; Big Blue unveiling". *USA Today*: p. 2B.

[16] "Kyocera QCP 6035 Smartphone Review" (http://www.palminfocenter.com/view_story.asp?ID=1707). Palminfocenter.com. 2001-03-16. . Retrieved 2011-09-07.

[17] Segan, Sascha (2010-03-23). "Kyocera Launches First Smartphone In Years | News & Opinion" (http://www.pcmag.com/article2/0,2817,2361664,00.asp). PCmag.com. . Retrieved 2011-09-07.

[18] "PDA Review: Ericsson R380 Smartphone" (http://www.geek.com/hwswrev/pda/ericr380/). *Geek.com*. . Retrieved 2011-12-15.

[19] "Symbian Device – The OS Evolution" (http://www.i-symbian.com/wp-content/uploads/2009/11/Symbian_Evolution.pdf) (PDF). *Independent Symbian Blog*. . Retrieved 2011-12-15.

[20] "Ericsson Introduces The New R380e" (http://www.mobilemag.com/2001/09/25/ericsson-introduces-the-new-r380e). *Mobile Magazine*. . Retrieved 2011-12-15.

[21] Brown, Bruce. "Ericsson R380 World, Review & Rating" (http://www.pcmag.com/article2/0,2817,40827,00.asp). *PCMag.com*. . Retrieved 2011-12-15.

[22] *Popular Science, December 1999* (http://books.google.com/books?id=8qSgh_Q-YOkC). *Google Books*. . Retrieved 2011-12-15.

[23] "Ericsson R380 PDA & Phone" (http://www.cellular.co.za/ericsson_r380.htm). *CellularOnline*. . Retrieved 2011-12-15.

[24] "Sony Ericsson P800" (http://www.allaboutsymbian.com/features/item/Sony_Ericsson_P800.php). *All About Symbian*. . Retrieved 2011-12-15.

[25] "Nokia N8 smartphone official site" (http://www.nokia.co.uk/gb-en/products/phone/n8-00/). . Retrieved 2011-12-15.

[26] "Nokia N8 review – best camera phone ever" (http://www.mobileburn.com/review.jsp?Id=11093). Mobileburn.com. 2010-10-05. . Retrieved 2011-12-15.

[27] "Mobile Choice Award: Best Sat-Nav" (http://www.mobilechoiceuk.com/News/mobile+choice+award:+Best+Sat-Nav/5274). Mobile Choice. 22 October 2010. . Retrieved 2011-12-15.

[28] "Nokia, Microsoft in pact to rival Apple, Google – Technology & Science" (http://www.cbc.ca/news/technology/story/2011/02/11/nokia-microsoft-smart-phone-apple-google.html). Associated Press. CBC.ca. 2011-02-11. . Retrieved 2011-12-15.

[29] Velazco, Chris (26 October 2011). "Nokia Debuts Their First Windows Phones: The Lumia 800 and Lumia 710" (http://techcrunch.com/2011/10/26/nokia-debuts-lumia-710-and-lumia-800/). *TechCruch*. . Retrieved 26 October 2011.

[30] "Better Living through Software: Microsoft Advances for the Home Highlighted at Consumer Electronics Show 2002: Microsoft showcases new and improved products – including "Freestyle," "Mira" and Ultimate TV at International CES 2002" (http://www.microsoft.com/presspass/features/2002/Jan02/01-08msces.mspx). Microsoft.com. 2002-01-08. . Retrieved 2011-09-07.
[31] Stephen H. Wildstrom (November 30, 2001). "Handspring's Breakthrough Hybrid" (http://www.businessweek.com/bwdaily/dnflash/nov2001/nf20011129_0157.htm). Businessweek.com. . Retrieved 2011-12-15.
[32] Kevin McLaughlin (December 17, 2009). "BlackBerry Users Call For RIM To Rethink Service" (http://www.crn.com/news/client-devices/222002587/blackberry-users-call-for-rim-to-rethink-service.htm). CRN.com. . Retrieved 2011-12-15.
[33] "iPhone in depth: the Ars review" (http://arstechnica.com/apple/reviews/2007/07/iphone-review.ars/6). *ArsTechnica*. Condé Nast. 9 July 2007. p. 6. . Retrieved 3 August 2010.
[34] "iPhone to Support Third-Party Web 2.0 Applications" (http://www.apple.com/pr/library/2007/06/11iphone.html). *Press Release*. Apple Inc.. June 11, 2007. . Retrieved December 15, 2008.
[35] "The iPhone is not a smartphone" (http://www.engadget.com/2007/01/09/the-iphone-is-not-a-smartphone/). Engadget.com. 9 January 2007. . Retrieved 11 July 2010.
[36] "Smartphones Can Replace These Everyday Items" (http://simpleorganizedlife.com/smartphones-can-replace-these-everyday-items/). Simple. Organized. Life.. October 10, 2011. . Retrieved 2011-12-15.
[37] "iPhone 3G on Sale Tomorrow" (http://www.apple.com/pr/library/2008/07/10iphone.html). *Press Release*. Apple Inc.. 2008-07-10. . Retrieved 2009-01-17.
[38] "iPhone App Store Downloads Top 10 Million in First Weekend" (http://www.apple.com/pr/library/2008/07/14iPhone-App-Store-Downloads-Top-10-Million-in-First-Weekend.html). *Press Release*. Apple Inc.. July 14, 2008. . Retrieved 2011-12-15.
[39] "Apple's App Store Downloads Top 1.5 Billion in First Year" (http://www.apple.com/pr/library/2009/07/14Apples-App-Store-Downloads-Top-1-5-Billion-in-First-Year.html). *Press Release*. Apple Inc.. July 14, 2009. . Retrieved 2011-12-15.
[40] "Apple's App Store Downloads Top 15 Billion" (http://www.apple.com/pr/library/2011/07/07Apples-App-Store-Downloads-Top-15-Billion.html). *Press Release*. Apple Inc.. July 7, 2011. . Retrieved 2011-12-15.
[41] "Apple now accepting iOS 4 apps, multitasking ahoy" (http://www.engadget.com/2010/06/11/apple-now-accepting-ios-4-apps-multitasking-ahoy/). Engadget.com. 11 June 2010. . Retrieved 2011-12-15.
[42] "Apple Presents iPhone 4" (http://www.apple.com/pr/library/2010/06/07Apple-Presents-iPhone-4.html). *Press Release*. Apple Inc.. 2010-06-07. . Retrieved 2011-07-05.
[43] "Liveblog: The Verizon iPhone" (http://voices.washingtonpost.com/fasterforward/2011/01/liveblog_the_verizon_iphone.html). *The Washington Post*. .
[44] Memmott, Mark (2011-01-11). "It's Official: Verizon Has The iPhone 4 : The Two-Way" (http://www.npr.org/blogs/thetwo-way/2011/01/11/132833078/its-official-verizon-has-iphone-4). NPR. . Retrieved 2011-09-07.
[45] Raice, Shayndi (January 12, 2011). "Verizon Unwraps iPhone" (http://online.wsj.com/article/SB10001424052748703791904576075681886276172.html?mod=googlenews_wsj). *The Wall Street Journal*. .
[46] Glenn Fleishman (2011-02-22). "Using the Personal Hotspot on your Verizon iPhone" (http://www.macworld.com/article/158058/2011/02/personal_hotspot_verizon.html). Macworld.com. . Retrieved 2011-03-12.
[47] "iOS 4.3 Software Update" (http://www.apple.com/ios/). Apple Inc.. . Retrieved 2011-03-12.
[48] Dan Moren (2011-03-11). "Hands on with iOS 4.3" (http://www.macworld.com/article/158483/2011/03/firstlook_43.html). Macworld.com. . Retrieved 2011-03-12.
[49] "Apple Launches iPhone 4S, iOS 5 & iCloud" (http://www.apple.com/pr/library/2011/10/04Apple-Launches-iPhone-4S-iOS-5-iCloud.html). Apple Inc.. October 4, 2011. . Retrieved 2011-12-15.
[50] "iPhone 4S Pre-Orders Top One Million in First 24 Hours" (http://www.apple.com/pr/library/2011/10/10iPhone-4S-Pre-Orders-Top-One-Million-in-First-24-Hours.html). Apple. . Retrieved 10 October 2011.
[51] Anderson, Ash. "iPhone 4S Sells 1 Million in Under 24 Hours" (http://www.keynoodle.com/iphone-4s-sells-1-million-in-under-24-hours/). *KeyNoodle*. . Retrieved 2011-12-15.
[52] "Apple's fall from grace" (http://www.apple.com/pr/library/2011/10/04Apple-Launches-iPhone-4S-iOS-5-iCloud.html). BGR.com. October 5, 2011. . Retrieved 2011-12-15.
[53] Michael E. Cohen (October 13, 2011). "iPhone 4S: A Very Palpable Hit" (http://tidbits.com/article/12554). tidbits.com. . Retrieved 2011-12-15.
[54] David Pogue (October 11, 2011). "New iPhone Conceals Sheer Magic" (http://www.nytimes.com/2011/10/12/technology/personaltech/iphone-4s-conceals-sheer-magic-pogue.html?_r=2&pagewanted=all). The New York Times. . Retrieved 2011-12-15.
[55] "Press Info – Apple to Launch iCloud on October 12" (http://www.apple.com/pr/library/2011/10/04Apple-to-Launch-iCloud-on-October-12.html). Apple. 2011-10-04. . Retrieved 2012-01-05.
[56] "Press Info – New Version of iOS Includes Notification Center, iMessage, Newsstand, Twitter Integration Among 200 New Features" (http://www.apple.com/pr/library/2011/06/06New-Version-of-iOS-Includes-Notification-Center-iMessage-Newsstand-Twitter-Integration-Among-200-New-Features.html). Apple. 2011-06-06. . Retrieved 2012-01-05.
[57] "Alliance Members" (http://www.openhandsetalliance.com/oha_members.html). *Open Handset Alliance*. . Retrieved 16 January 2011.
[58] "Weighing Nexus over iPhone – a practical review for the everyday user!" (http://techietrick.blogspot.com/2010/01/weighing-nexus-over-iphone-practical.html). Techietrick.blogspot.com. 2010-01-06. . Retrieved 2011-09-07.

[59] "Nexus One gets a software update, enables multitouch (updated with video!)" (http://www.engadget.com/2010/02/02/nexus-one-gets-a-software-update-enables-multitouch). Engadget.com. . Retrieved 2011-09-07.
[60] Tarmo Virki (2011-02-14). "Sony takes gaming console war to phones" (http://in.reuters.com/article/2011/02/13/idINIndia-54865020110213). Reuters. .
[61] "HTC EVO 3D" (http://web.archive.org/web/20110511105547/http://www.htc.com/www/product/evo3d/overview.html). htc.com. Archived from the original (http://www.htc.com/www/product/evo3d/overview.html) on 2011-05-11. . Retrieved 2011-12-15.
[62] Savov, Vlad. "HTC Evo 3D Launches June 24th" (http://www.engadget.com/2011/06/06/htc-evo-3d-launches-on-june-24th-for-200-joined-by-evo-view-4g/). Engadget.com. . Retrieved 7 June 2011.
[63] Ed Hansberry (11 November 2009). "Samsung Bailing on Windows Mobile" (http://www.informationweek.com/blog/main/archives/2009/11/samsung_bailing.html). *InformationWeek*. .
[64] "Samsung to Discard Windows Phone" (http://www.telecomskorea.com/market-8281.html). *Telecoms Korea*. 9 November 2009. .
[65] "Samsung Wave, first Bada smartphone hits the market" (http://www.bada.com/samsung-wave-first-bada-smartphone-hits-the-market/). *Bada*. 24 May 2010. . Retrieved 3 February 2011.
[66] "BadaWave" (http://badawave.com/). BadaWave. . Retrieved 2012-01-05.
[67] "Samsung Waves away a million" (http://www.theinquirer.net/inquirer/news/1722287/samsung-waves-away-million). The Inquirer. 13 July 2010. .
[68] "Android increases smart phone market leadership with 35% share" (http://www.canalys.com/newsroom/android-increases-smart-phone-market-leadership-35-share). Canalys.com. 2011-05-04. . Retrieved 2011-09-07.
[69] "Samsung Bada shipments up 355% to 4.5 million units in Q2 2011 | asymco news | PG.Biz" (http://www.pocketgamer.co.uk/r/PG.Biz/asymco+news/news.asp?c=32049). Pocket Gamer. . Retrieved 2011-09-07.
[70] Menezes, Gary (2010-09-11). "Symbian OS, Now Fully Open Source" (http://www.watblog.com/2010/02/06/symbian-os-now-fully-open-source). Watblog.com. . Retrieved 2011-09-07.
[71] "Nokia N9 Smart Phone Review" (http://wontek.com/smart-phones/nokia/nokia-n9-smart-phone-review). Wontek.com. 2011-01-01. . Retrieved 2011-09-07.
[72] "McAfee: Android malware surges 76%, iPhone untouched" (http://www.electronista.com/articles/11/08/23/mcafee.shows.android.facing.huge.spike.in.malware/). Electronista.com. . Retrieved 2012-01-05.
[73] Dalrymple, Jim (2011-11-16). "Android sees a 472% increase in malware since July" (http://www.loopinsight.com/2011/11/16/android-sees-a-472-increase-in-malware-since-july/). Loopinsight.com. . Retrieved 2012-01-05.
[74] Mobile Malware Development Continues To Rise, Android Leads The Way (http://globalthreatcenter.com/?p=2492).
[75] "The Mother Of All Android Malware Has Arrived" (http://www.androidpolice.com/2011/03/01/the-mother-of-all-android-malware-has-arrived-stolen-apps-released-to-the-market-that-root-your-phone-steal-your-data-and-open-backdoor/). *Android Police*. March 6, 2011. .
[76] Perez, Sarah (2009-02-12). "Android Vulnerability So Dangerous, Owners Warned Not to Use Phone's Web Browser" (http://www.readwriteweb.com/archives/android_vulnerability_so_dangerous_shouldnt_use_web_browser.php). Readwriteweb.com. . Retrieved 2011-08-08.
[77] "Lookout, Retrevo warn of growing Android malware epidemic, note Apple's iOS is far safer" (http://www.appleinsider.com/articles/11/08/03/lookout_retrevio_warn_of_growing_android_malware_epidemic_note_apples_ios_is_far_safer.html). Appleinsider.com. 2011-08-03. . Retrieved 2012-01-05.
[78] "First SMS Trojan detected for smartphones running Android" (http://www.kaspersky.com/news?id=207576158). Kaspersky Lab. . Retrieved 2010-10-18.
[79] "Apple's iOS unaffected by malware as Android exploits surge 76%" (http://www.appleinsider.com/articles/11/08/24/apples_ios_unaffected_by_malware_as_android_exploits_surge_76.html). Appleinsider.com. 2011-08-24. . Retrieved 2012-01-05.
[80] "Security researcher finds code signing flaw that opens door for iOS malware" (http://www.appleinsider.com/articles/11/11/07/newly_found_code_signing_flaw_allows_for_ios_malware.html). Appleinsider.com. 2011-11-07. . Retrieved 2012-01-05.
[81] Apple's iOS more secure than Google's Android, says Symantec (http://www.appleinsider.com/articles/11/06/28/apples_ios_more_secure_than_googles_android_says_symantec.html). Appleinsider.com (2011-06-28). Retrieved on 2012-08-09.
[82] Android Orphans: Visualizing a Sad History of Support (http://theunderstatement.com/post/11982112928/android-orphans-visualizing-a-sad-history-of-support). the understatement (2011-10-26). Retrieved on 2012-08-09.
[83] Apple Inc.. "Distribute your App – iOS Developer Program – Apple Developer" (http://developer.apple.com/programs/ios/distribute.html). Developer.apple.com. . Retrieved 2012-01-05.
[84] "Apple's rivals battle for iOS scraps as app market sales grow to $2.2 billion" (http://www.appleinsider.com/articles/11/02/18/rim_nokia_and_googles_android_battle_for_apples_ios_scraps_as_app_market_sales_grow_to_2_2_billion.html). Appleinsider.com. 2011-02-18. . Retrieved 2012-01-05.
[85] "Google Android has double the number of free apps than Apple's App Store" (http://techcrunch.com/2010/07/05/distimo-june-2010/). Distimo. 15 July 2009. . Retrieved 15 July 2009.
[86] Lance Whitney, CNET. " Nearly 1 in 5 smartphone owners use check-in services (http://news.cnet.com/8301-1023_3-20062640-93.html?part=rss&subj=news&tag=2547-1_3-0-20)." May 13, 2011. Retrieved May 13, 2011.
[87] Second Screen change the way we watch TV (http://www.good.is/post/double-the-glow-will-second-screen-apps-change-the-way-we-watch-tv/). Good.is (2011-08-12). Retrieved on 2012-08-09.

[88] explosion of second screen apps (http://www.adweek.com/news/technology/second-screen-apps-explode-132237). Adweek.com. Retrieved on 2012-08-09.
[89] "Smart phones: how to stay clever in downturn" (http://www.deloitte.co.uk/TMTPredictions/telecommunications/Smartphones-clever-in-downturn.cfm). *Deloitte Telecommunications Predictions*. .
[90] "Android Phones Steal Market Share" (http://bmighty.informationweek.com/mobile/showArticle.jhtml?articleID=224201881). .
[91] "100 Million Club – H1 2010" (http://www.visionmobile.com/blog/2010/10/smart-feature-phones-the-unbalanced-equation-100-million-club-series/). .
[92] "Mobile Phone Development » Blog Archive » Gartner Q3 2010" (http://www.mobilephonedevelopment.com/archives/1149). Mobilephonedevelopment.com. 2010-11-10. . Retrieved 2011-09-07.
[93] Stan Schroeder (2011-02-10). "Gartner: Symbian Is Still the Number One Smartphone Platform" (http://mashable.com/2011/02/10/symbian-number-one-gartner-report/). Mashable.com. . Retrieved 2011-09-07.
[94] "Gartner Says Worldwide Mobile Device Sales to End Users Reached 1.6 Billion Units in 2010; Smartphone Sales Grew 72 Percent in 2010" (http://www.gartner.com/it/page.jsp?id=1543014). Gartner.com. 2011-02-11. . Retrieved 2011-09-07.
[95] "Olswang Predicts Mobile's Impact on TV: The Convergent Revolution" (http://www.tvgenius.net/blog/2011/03/17/mobile-tv-convergence/). Tvgenius.net. 2011-03-17. . Retrieved 2011-09-07.
[96] "Berg: Smartphone shipments grew 74% in 2010" (http://www.bgr.com/2011/03/10/berg-smartphone-shipments-grew-74-in-2010/). *Boy Genius Report*. March 10, 2011. .
[97] Generation App: 62% of Mobile Users 25–34 own Smartphones | Nielsen Wire (http://blog.nielsen.com/nielsenwire/?p=29786).
[98] Market Research | Consumer Market Research – NPD – As Smartphone Prices Fall, Retailers Are Leaving Money on the Table, According to The NPD Group (http://www.npdgroup.com/wps/portal/npd/us/news/pressreleases/pr_111114a). Npdgroup.com (2011-11-14). Retrieved on 2012-08-09.
[99] Poeter, Damon. (2011-11-05) Apple, With 4 Percent of Handset Market, Captures 52 Percent of Profits (http://www.pcmag.com/article2/0,2817,2395951,00.asp#fbid=SLgcdJun7IU). Pcmag.com. Retrieved on 2012-08-09.
[100] "Smartphones killing point-and-shoots, now take almost 1/3 of photos" (http://gigaom.com/2011/12/22/smartphones-killing-point-and-shoots-now-take-almost-13-of-photos/). . Retrieved December 25, 2011.
[101] Strategy Analytics: Samsung Becomes World's Number One Smartphone Vendor in Q3 2011 (http://www.marketwatch.com/story/strategy-analytics-samsung-becomes-worlds-number-one-smartphone-vendor-in-q3-2011-2011-10-27). MarketWatch. Retrieved on 2012-08-09.
[102] Gartner Says Sales of Mobile Devices in Second Quarter of 2011 Grew 16.5 Percent Year-on-Year; Smartphone Sales Grew 74 Percent (http://www.gartner.com/it/page.jsp?id=1764714). Gartner.com. Retrieved on 2012-08-09.
[103] "Smartphone Sales Will Hit 420 Million In 2011, To Take 28 Percent Of The Total Phone Market" (http://techcrunch.com/2011/07/27/smartphone-sales-will-hit-420-million-in-2011-to-take-28-percent-of-the-total-phone-market/). TechCrunch. 27 July 2011. .
[104] Epstein, Zach. (2011-07-29) Apple's iPhone accounted for 66% of Q2 smartphone profit among top vendors (http://www.bgr.com/2011/07/29/apples-iphone-accounted-for-66-of-q2-smartphone-profit-among-top-vendors/). Bgr.com. Retrieved on 2012-08-09.
[105] "SA agrees: Apple now top smartphone vendor in the world with 140% growth" (http://www.bgr.com/2011/07/29/sa-agrees-apple-now-top-smartphone-vendor-in-the-world-with-240-growth/). BGR. 29 July 2011. .
[106] "Nokia reports Q2 results: Sells 88.5 million devices and declares net loss of 368 million euros" (http://mobilesyrup.com/2011/07/21/nokia-reports-q2-results-sells-88-5-million-devices-and-declares-net-loss-of-368-million-euros/). Mobilesyrup.com. 2011-07-21. . Retrieved 2011-09-07.
[107] "In U.S. Smartphone Market, Android is Top Operating System, Apple is Top Manufacturer" (http://blog.nielsen.com/nielsenwire/?p=28516). Nielsen. 28 July 2011. . Retrieved 5 September 2011.
[108] Isaac, Mike (28 July 2011). "Android Still Dominates Phones, But What About the Rest of Mobile?" (http://www.wired.com/gadgetlab/2011/07/android-ios-platform-share/all/1). Wired. . Retrieved 5 September 2011.
[109] iPhone 4 remains top-selling US smartphone despite growing iPhone 5 hype (http://www.appleinsider.com/articles/11/09/06/iphone_4_remains_top_selling_us_smartphone_despite_growing_iphone_5_hype.html). Appleinsider.com (2011-09-06). Retrieved on 2012-08-09.
[110] Previous-gen Apple iPad, iPhone 3GS often outsell new Android devices (http://www.appleinsider.com/articles/11/05/09/previous_gen_apple_ipad_iphone_3gs_often_outsell_new_android_devices.html). Appleinsider.com (2011-05-09). Retrieved on 2012-08-09.
[111] "Galaxy phones power Samsung to record $5.2 billion profit" (http://money.msn.com/business-news/article.aspx?feed=OBR&date=20120427&id=15039578). April 27, 2012. .
[112] Ever-Popular iPhone Named Top Smartphone. Again (http://www.wired.com/gadgetlab/2011/09/iphone-tops-survey-sixth-time/all/1).
[113] Wireless Consumer Smartphone Ratings (Volume 1) (http://www.jdpower.com/Electronics/ratings/wireless-consumer-smartphone-ratings-(volume-1)/). Jdpower.com. Retrieved on 2012-08-09.
[114] Wireless Consumer Smartphone Ratings (Volume 2) (http://www.jdpower.com/Electronics/ratings/wireless-consumer-smartphone-ratings-(volume-2)/).
[115] 2011 U.S. Wireless Handset Customer Satisfaction Studies—Vol. 2 (http://www.jdpower.com/news/pressRelease.aspx?ID=2011146)
[116] 2011 Wireless Smartphone and Traditional Mobile Phone Satisfaction Studies-Vol. 1 (http://www.jdpower.com/news/pressrelease.aspx?ID=2011030)

[117] 2010 U.S. Wireless Smartphone and Traditional Mobile Phone Customer Satisfaction Studies-Volume 1 (http://businesscenter.jdpower.com/news/pressrelease.aspx?ID=2010039). Businesscenter.jdpower.com (2010-04-01). Retrieved on 2012-08-09.
[118] 2009 Wireless Consumer Smartphone and Traditional Mobile Phone Customer Satisfaction Studies (http://businesscenter.jdpower.com/news/pressrelease.aspx?ID=2009082). Businesscenter.jdpower.com (2009-04-30). Retrieved on 2012-08-09.
[119] Tarmo Virki (2011-01-31). "Google topples Symbian from smart phones top spot" (http://www.theglobeandmail.com/news/technology/mobile-technology/google-topples-symbian-from-smartphones-top-spot/article1888557/?cmpid=rss1). Toronto: Theglobeandmail.com. . Retrieved 2011-09-07.
[120] Virki, Tarmo (2011-05-04). "Android became clear smartphone leader in first quarter: Canalys" (http://www.reuters.com/article/2011/05/04/us-smartphones-research-idUSTRE7434VV20110504). Reuters.com. . Retrieved 2011-09-07.
[121] "Android Takes Over in UK" (http://www.bestsmartphone.com/2011/11/02/android-takes-over-in-uk/). bestsmartphone.com. 2011-11-02. .
[122] "Proof of the iPhone's Dependence on Carrier Subsidies" (http://www.tech-thoughts.net/2012/05/proof-of-iphones-dependence-on-carrier.html). tech-thoughts.net. 2012-05-29. .
[123] Statistics Made Visual (http://www.mobilestatistics.com/mobile-statistics/). Mobile Statistics. Retrieved on 2012-08-09.
[124] Gartner Says Worldwide Smartphone Sales Reached Its Lowest Growth Rate With 3.7 Per Cent Increase in Fourth Quarter of 2008 (http://www.gartner.com/it/page.jsp?id=910112). Gartner.com. Retrieved on 2012-08-09.
[125] Gartner Says Worldwide Mobile Phone Sales to End Users Grew 8 Per Cent in Fourth Quarter 2009; Market Remained Flat in 2009 (http://www.gartner.com/it/page.jsp?id=1306513). Gartner.com. Retrieved on 2012-08-09.
[126] Gartner Says Worldwide Mobile Device Sales to End Users Reached 1.6 Billion Units in 2010; Smartphone Sales Grew 72 Percent in 2010 (http://www.gartner.com/it/page.jsp?id=1543014). Gartner.com. Retrieved on 2012-08-09.
[127] Statistics Made Visual (http://www.mobilestatistics.com/mobile-statistics/). Mobile Statistics. Retrieved on 2012-08-09.
[128] Gartner Says Worldwide Sales of Mobile Phones Declined 2 Percent in First Quarter of 2012; Previous Year-over-Year Decline Occurred in Second Quarter of 2009 (http://www.gartner.com/it/page.jsp?id=2017015). Gartner.com. Retrieved on 2012-08-09.
[129] iPass : Mobile Workforce Report (http://mobile-workforce-project.ipass.com/).
[130] iPhone pushes past BlackBerry to top enterprise phone ranks (http://www.appleinsider.com/articles/11/11/16/iphone_pushes_past_blackberry_to_top_enterprise_phone_ranks.html). Appleinsider.com (2011-11-16). Retrieved on 2012-08-09.
[131] In the US Market, iPhone Outperforms Other Mobile Platforms in User Loyalty by a Wide Margin, Android is Second, Blackberry Fourth (http://www.zokem.com/2011/01/in-the-us-market-iphone-outperforms-other-mobile-platforms-in-user-loyalty-by-a-wide-margin-android-is-second-blackberry-fourth/). Zokem.com (2011-01-18). Retrieved on 2012-08-09.
[132] Clint Boulton (2011-01-20). iPhone Whips Android, BlackBerry in User Loyalty: Zokem (http://www.eweek.com/c/a/Mobile-and-Wireless/iPhone-Whips-Android-Blackberry-in-User-Loyalty-Zokem-445681/). eweek.com
[133] Europe's mobile operators look to international opportunities and LTE for future growth (http://www.telecomsmarketresearch.com/resources/Mobile_Market_Europe.shtml). telecomsmarketresearch.com
[134] "The magic moment:Smartphones now half of all U.S. mobiles" (http://venturebeat.com/2012/03/29/the-magic-moment-smartphones-now-half-of-all-u-s-mobiles/). venturebeat.com (2012-03-29).

Android_(operating_system)

Android

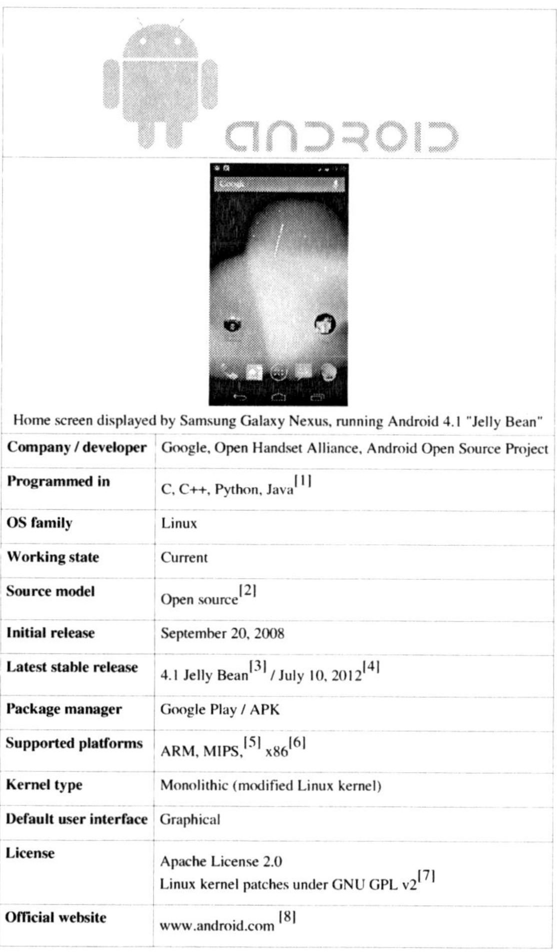

Home screen displayed by Samsung Galaxy Nexus, running Android 4.1 "Jelly Bean"

Company / developer	Google, Open Handset Alliance, Android Open Source Project
Programmed in	C, C++, Python, Java[1]
OS family	Linux
Working state	Current
Source model	Open source[2]
Initial release	September 20, 2008
Latest stable release	4.1 Jelly Bean[3] / July 10, 2012[4]
Package manager	Google Play / APK
Supported platforms	ARM, MIPS,[5] x86[6]
Kernel type	Monolithic (modified Linux kernel)
Default user interface	Graphical
License	Apache License 2.0 Linux kernel patches under GNU GPL v2[7]
Official website	www.android.com [8]

Android is a Linux-based operating system primarily designed for mobile devices such as smartphones and tablet computers utilizing ARM processors. A secondary target for the light weight OS is embedded systems such as networking equipment, smart TV systems including set top boxes and built in systems and various devices as varied as house hold appliances and wrist watches. Most embedded applications are for ARM based devices but notably Google's Google TV devices use Intel chips with the x86 version of Android. The x86 processor architecture is also utilized, to a lesser extent, in traditional personal computer applications most notably with netbooks and, rarely, laptops and desktops. It is developed by the Open Handset Alliance, led by Google.[2]

Google financially backed the initial developer of the software, Android Inc., and later purchased it in 2005.[9] The unveiling of the Android distribution in 2007 was announced with the founding of the Open Handset Alliance, a consortium of 86 hardware, software, and telecommunication companies devoted to advancing open standards for mobile devices.[10] Google releases the Android code as open-source, under the Apache License.[11] The Android Open Source Project (AOSP) is tasked with the maintenance and further development of Android.[12]

Android has a large community of developers writing applications ("apps") that extend the functionality of the devices. Developers write primarily in a customized version of Java.[13] Apps can be downloaded from third-party sites or through online stores such as Google Play (formerly *Android Market*), the app store run by Google. In June 2012, there were more than 600,000 apps available for Android, and the estimated number of applications downloaded from Google Play was 20 billion.[14]

Android became the world's leading smartphone platform at the end of 2010.[15] For the first quarter of 2012, Android had a 59% smartphone market share worldwide.[16] At the half of 2012, there were 400 million devices activated and 1 million activations per day.[17] Analysts point to the advantage to Android of being a multi-channel, multi-carrier OS.[18]

History

Foundation

Android, Inc. was founded in Palo Alto, California, United States in October 2003 by Andy Rubin (co-founder of Danger),[19] Rich Miner (co-founder of Wildfire Communications, Inc.),[20] Nick Sears[21] (once VP at T-Mobile),[22] and Chris White (headed design and interface development at WebTV)[9] to develop, in Rubin's words "...smarter mobile devices that are more aware of its owner's location and preferences".[9] Despite the obvious past accomplishments of the founders and early employees, Android Inc. operated secretly, revealing only that it was working on software for mobile phones.[9] That same year, Rubin ran out of money. Steve Perlman, a close friend of Rubin, brought him $10,000 in cash in an envelope and refused a stake in the company.[23]

Google acquisition

Google acquired Android Inc. on August 17, 2005, making Android Inc. a wholly owned subsidiary of Google. Key employees of Android Inc., including Andy Rubin, Rich Miner and Chris White, stayed at the company after the acquisition.[9] Not much was known about Android Inc. at the time of the acquisition, but many assumed that Google was planning to enter the mobile phone market with this move.[9]

At Google, the team led by Rubin developed a mobile device platform powered by the Linux kernel. Google marketed the platform to handset makers and carriers on the promise of providing a flexible, upgradable system. Google had lined up a series of hardware component and software partners and signaled to carriers that it was open to various degrees of cooperation on their part.[24][25][26]

Speculation about Google's intention to enter the mobile communications market continued to build through December 2006.[27] Reports from the BBC and *The Wall Street Journal* noted that Google wanted its search and applications on mobile phones and it was working hard to deliver that. Print and online media outlets soon reported rumors that Google was developing a Google-branded handset. Some speculated that as Google was defining technical specifications, it was showing prototypes to cell phone manufacturers and network operators.

In September 2007, *InformationWeek* covered an Evalueserve study reporting that Google had filed several patent applications in the area of mobile telephony.[28][29]

Open Handset Alliance

On November 5, 2007, the Open Handset Alliance, a consortium of several companies which include Broadcom Corporation, Google, HTC, Intel, LG, Marvell Technology Group, Motorola, Nvidia, Qualcomm, Samsung Electronics, Sprint Nextel, T-Mobile and Texas Instruments unveiled itself. The goal of the Open Handset Alliance is to develop open standards for mobile devices.[10] On the same day, the Open Handset Alliance also unveiled their first product, Android, a mobile device platform built on the Linux kernel version 2.6.[10]

On December 9, 2008, 14 new members joined, including ARM Holdings, Atheros Communications, Asustek Computer Inc, Garmin Ltd, Huawei Technologies, PacketVideo, Softbank, Sony Ericsson, Toshiba Corp, and Vodafone Group Plc.[30][31]

Android Open Source Project

The Android Open Source Project (AOSP) [32] is led by Google, and is tasked with the maintenance and development of Android.[33] According to the project "The goal of the Android Open Source Project is to create a successful real-world product that improves the mobile experience for end users."[34] AOSP also maintains the *Android Compatibility Program*, defining an "Android compatible" device "as one that can run any application written by third-party developers using the Android SDK and NDK", to prevent incompatible Android implementations.[34] The compatibility program is also optional and free of charge, with the *Compatibility Test Suite* also free and open-source.[35]

Version history

Android has been updated frequently since the original release of "Astro", with each fixing bugs and adding new features. Each version after Astro and Bender is named in alphabetical order after a dessert or sweet treat, with 1.5 "Cupcake" being the first and every update since following this naming convention.[36]

From left to right: HTC Dream (G1), Nexus One, Nexus S, Galaxy Nexus

List of Android version code names:

- Cupcake
- Donut
- Eclair
- Froyo
- Gingerbread
- Honeycomb
- Ice Cream Sandwich
- Jelly Bean

Design

Android consists of a kernel based on the Linux kernel 2.6, with middleware, libraries and APIs written in C and application software running on an application framework which includes Java-compatible libraries based on Apache Harmony. Android uses the Dalvik virtual machine with just-in-time compilation to run Dalvik dex-code (Dalvik Executable), which is usually translated from Java bytecode.[37]

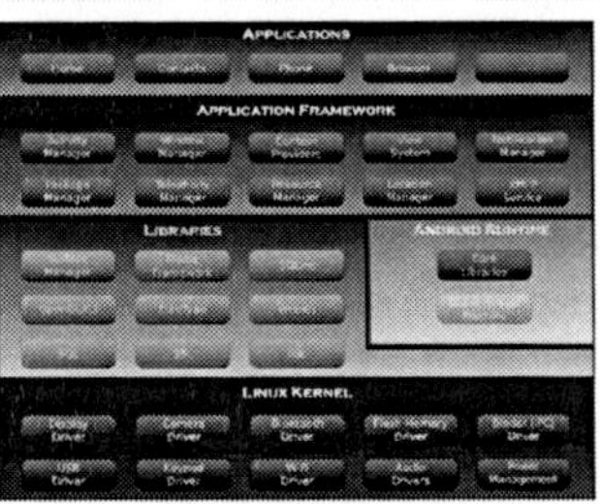

Architecture diagram

The main hardware platform for Android is the ARM architecture. There is support for x86 from the Android x86 project,[6] and Google TV uses a special x86 version of Android.

Linux

Android's kernel is based on the Linux kernel and has further architecture changes by Google outside the typical Linux kernel development cycle.[38] Android does not have a native X Window System by default nor does it support the full set of standard GNU libraries, and this makes it difficult to port existing Linux applications or libraries to Android.[39]

Certain features that Google contributed back to the Linux kernel, notably a power management feature called wakelocks, were rejected by mainline kernel developers, partly because kernel maintainers felt that Google did not show any intent to maintain their own code.[40][41][42] Even though Google announced in April 2010 that they would hire two employees to work with the Linux kernel community,[43] Greg Kroah-Hartman, the current Linux kernel maintainer for the -stable branch, said in December 2010 that he was concerned that Google was no longer trying to get their code changes included in mainstream Linux.[41] Some Google Android developers hinted that "the Android team was getting fed up with the process", because they were a small team and had more urgent work to do on Android.[44]

However, in September 2010, Linux kernel developer Rafael J. Wysocki added a patch that improved the mainline Linux wakeup events framework. He said that Android device drivers that use wakelocks can now be easily merged into mainline Linux, but that Android's opportunistic suspend features should not be included in the mainline kernel.[45][46] In August 2011, Linus Torvalds said that "eventually Android and Linux would come back to a common kernel, but it will probably not be for four to five years".[47]

In December 2011, Greg Kroah-Hartman announced the start of the Android Mainlining Project, which aims to put some Android drivers, patches and features back into the Linux kernel, starting in Linux 3.3.[48] further integration being expected for Linux Kernel 3.4.[49]

Features

Current features and specifications:[50][51][52]

The Android Emulator default home screen (v1.5, also known as "Cupcake")

Handset layouts

The platform is adaptable to larger, VGA, 2D graphics library, 3D graphics library based on OpenGL ES 2.0 specifications, and traditional smartphone layouts.

Storage

SQLite, a lightweight relational database, is used for data storage purposes.

Connectivity

Android supports connectivity technologies including GSM/EDGE, IDEN, CDMA, EV-DO, UMTS, Bluetooth, Wi-Fi, LTE, NFC and WiMAX.

Messaging

SMS and MMS are available forms of messaging, including threaded text messaging and Android Cloud To Device Messaging (C2DM) and now enhanced version of C2DM, Android Google Cloud Messaging (GCM) is also a part of Android Push Messaging service.

Multiple language support

Android supports multiple languages.[53]

Web browser

The web browser available in Android is based on the open-source WebKit layout engine, coupled with Chrome's V8 JavaScript engine. The browser scores 100/100 on the Acid3 test on Android 4.0.

Java support

While most Android applications are written in Java, there is no Java Virtual Machine in the platform and Java byte code is not executed. Java classes are compiled into Dalvik executables and run on Dalvik, a specialized virtual machine designed specifically for Android and optimized for battery-powered mobile devices with limited memory and CPU. J2ME support can be provided via third-party applications.

Media support

Android supports the following audio/video/still media formats: WebM, H.263, H.264 (in 3GP or MP4 container), MPEG-4 SP, AMR, AMR-WB (in 3GP container), AAC, HE-AAC (in MP4 or 3GP container), MP3, MIDI, Ogg Vorbis, FLAC, WAV, JPEG, PNG, GIF, BMP, WebP.[52]

Streaming media support

RTP/RTSP streaming (3GPP PSS, ISMA), HTML progressive download (HTML5 <video> tag). Adobe Flash Streaming (RTMP) and HTTP Dynamic Streaming are supported by the Flash plugin.[54] Apple HTTP Live Streaming is supported by RealPlayer for Android,[55] and by the operating system in Android 3.0 (Honeycomb).[56]

Additional hardware support

Android can use video/still cameras, touchscreens, GPS, accelerometers, gyroscopes, barometers, magnetometers, dedicated gaming controls, proximity and pressure sensors, thermometers, accelerated 2D bit blits (with hardware orientation, scaling, pixel format conversion) and accelerated 3D graphics.

Multi-touch

Android has native support for multi-touch which was initially made available in handsets such as the HTC Hero. The feature was originally disabled at the kernel level (possibly to avoid infringing Apple's patents on touch-screen technology at the time).[57] Google has since released an update for the Nexus One and the Motorola Droid which enables multi-touch natively.[58]

Bluetooth

Supports A2DP, AVRCP, sending files (OPP), accessing the phone book (PBAP), voice dialing and sending contacts between phones. Keyboard, mouse and joystick (HID) support is available in Android 3.1+, and in earlier versions through manufacturer customizations and third-party applications.[59]

Video calling

Android does not support native video calling, but some handsets have a customized version of the operating system that supports it, either via the UMTS network (like the Samsung Galaxy S) or over IP. Video calling through Google Talk is available in Android 2.3.4 and later. Gingerbread allows Nexus S to place Internet calls with a SIP account. This allows for enhanced VoIP dialing to other SIP accounts and even phone numbers. Skype 2.1 offers video calling in Android 2.3, including front camera support.

Multitasking

Multitasking of applications, with unique handling of memory allocation, is available.[60]

Voice based features

Google search through voice has been available since initial release.[61] Voice actions for calling, texting, navigation, etc. are supported on Android 2.2 onwards.[62]

Tethering

Android supports tethering, which allows a phone to be used as a wireless/wired Wi-Fi hotspot. Before Android 2.2 this was supported by third-party applications or manufacturer customizations.[63]

Screen capture

Android supports capturing a screenshot by pressing the power and volume-down buttons at the same time.[64] Prior to Android 4.0, the only methods of capturing a screenshot were through manufacturer and third-party customizations or otherwise by using a PC connection (DDMS developer's tool). These alternative methods are still available with the latest Android.

External storage

Most Android devices include microSD slot and can read microSD cards formatted with FAT32, Ext3 or Ext4 file system. To allow use of high-capacity storage media such as USB flash drives and USB HDDs, many Android tablets also include USB 'A' receptacle. Storage formatted with FAT32 is handled by Linux Kernel VFAT driver, while 3rd party solutions are required to handle other popular file systems such as NTFS, HFS Plus and exFAT.

Uses

While Android is designed primarily for smartphones and tablets, the open and customizable nature of the operating system allows it to be used on other electronics, including laptops and netbooks, smartbooks,[65] ebook readers,[66] and smart TVs (Google TV). Further, the OS has seen niche applications on wristwatches,[67] headphones,[68] car CD and DVD players,[69] smart glasses (Project Glass), refrigerators, vehicle satnav systems, home automation systems, games consoles, mirrors,[70] cameras,[71] portable media players[72] landlines,[73] and treadmills.[74]

The first commercially available phone to run Android was the HTC Dream, released on October 22, 2008.[75] In early 2010 Google collaborated with HTC to launch its flagship[76] Android device, the Nexus One. This was followed later in 2010 with the Samsung-made Nexus S and in 2011 with the Galaxy Nexus.

iOS and Android 2.3.3 'Gingerbread' may be set up to dual boot on a jailbroken iPhone or iPod Touch with the help of OpeniBoot and iDroid.[77][78]

In December 2011 it was announced the Pentagon has officially approved Android for use by its personnel.[79][80][81]

Applications

Applications are usually developed in the Java language using the Android Software Development Kit, but other development tools are available, including a Native Development Kit for applications or extensions in C or C++, Google App Inventor, a visual environment for novice programmers and various cross platform mobile web applications frameworks.

Applications can be acquired by end-users either through a store such as Google Play or the Amazon Appstore, or by downloading and installing the application's APK file from a third-party site.[82]

Google Play

Google Play is an online software store developed by Google for Android devices. An application program ("app") called "Play Store" is preinstalled on most Android devices and allows users to browse and download apps published by third-party developers, hosted on Google Play. As of June 2012, there were more than 600,000 apps available for Android, and the estimated number of applications downloaded from the Play Store exceeded 20 billion[14]. The operating system itself is installed on 400 million total devices.[17]

Only devices that comply with Google's compatibility requirements are allowed to preinstall and access the Play Store.[83] The app filters the list of available applications to those that are compatible with the user's device, and developers may restrict their applications to particular carriers or countries for business reasons.[84]

Google offers many free applications in the Play Store including Google Voice, Google Goggles, Gesture Search, Google Translate, Google Shopper, Listen and My Tracks. In August 2010, Google launched "Voice Actions for Android",[85] which allows users to search, write messages, and initiate calls by voice.

Security

Android applications run in a sandbox, an isolated area of the operating system that does not have access to the rest of the system's resources, unless access permissions are granted by the user when the application is installed. Before installing an application, the Play Store displays all required permissions. A game may need to enable vibration, for example, but should not need to read messages or access the phonebook. After reviewing these permissions, the user can decide whether to install the application.[86] The sandboxing and permissions system weakens the impact of vulnerabilities and bugs in applications, but developer confusion and limited documentation has resulted in applications routinely requesting unnecessary permissions, reducing its effectiveness.[87] The complexity of inter-application communication implies Android may have opportunities to run unauthorized code.[88]

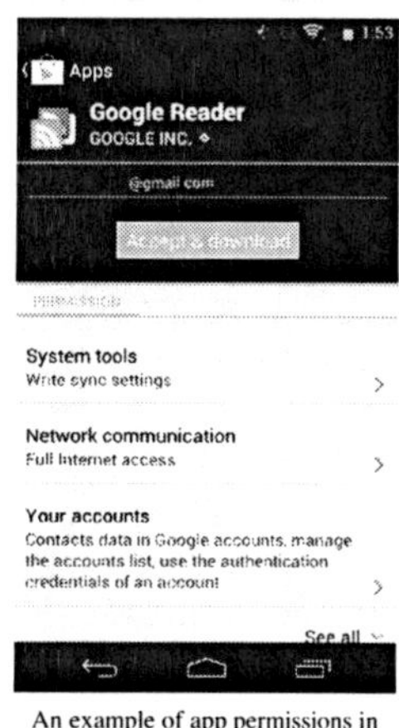

An example of app permissions in the Play Store.

Several security firms have released antivirus software for Android devices, in particular, Lookout Mobile Security,[89] AVG Technologies,[90] Avast!,[91] F-Secure,[92] Kaspersky,[93] McAfee[94] and Symantec.[95] This software is ineffective as sandboxing also applies to such applications, limiting their ability to scan the deeper system for threats.[96]

A useful type of security applications program and service, often described as "Find My Phone", is available for Android, as well as for Microsoft Windows Phone and for Apple iPhone, whereby a registered user can find the approximate location of the phone, if switched on, over the Internet. This helps to locate lost or stolen phones. At least one of these can be installed on a phone after it has gone missing.[97]

Privacy

Android smartphones have the ability to report the location of Wi-Fi access points, encountered as phone users move around, to build databases containing the physical locations of hundreds of millions of such access points. These databases form electronic maps to locate smartphones, allowing them to run apps like Foursquare, Latitude, Places, and to deliver location-based ads.[98]

Third party monitoring software such as TaintDroid,[99] an academic research-funded project, can, in some cases, detect when personal information is being sent from applications to remote servers.[100]

In March 2012 it was revealed that Android Apps can copy photos without explicit user permission,[101] Google responded they "originally designed the Android photos file system similar to those of other computing platforms like Windows and Mac OS. [...] we're taking another look at this and considering adding a permission for apps to access images. We've always had policies in place to remove any apps *[on Google Play]* that improperly access your data."[102]

Marketing

The Android logotype was designed along with the Droid font family by Ascender Corporation,[103] the robot icon was designed by Irina Blok.[104]

Android Green is the color of the Android Robot that represents the Android operating system. The print color is PMS 376C and the RGB color value in hexadecimal is #A4C639, as specified by the Android Brand Guidelines.[105] The custom typeface of Android is called Norad (cf. NORAD). It is only used in the text logo.[105]

Market share

Period	Worldwide smartphones market	U.S. smartphone market	Global devices	Activations per day	U.S devices	Source
Q2 2009	2.8%			12,100		Canalys[106]
Q3 2009	4%[107]	8%[108]		18,000		
Q4 2009	8.7%			51,100		Canalys[15]
2009		9.7%				North America, Canalys[109]
February 2010					4,09 million	9% of 45.4 million U.S. smartphones, ComScore [110]
Q1 2010		28 %				NPD Group[111] Android outsold Apple's iPhone in the U.S.
May 2010				100,000[112]		
June 2010		33%[108]		160,000[112]		
Q3 2010	25.3%	43.6%[113]		223,000		Gartner[114]
September 2010					12.6 million	21.4% of the 58.7 million U.S. smartphones[115]

Q4 2010	32.9%			362,000		Canalys.[15]
February 2011					23.8 million	comscore[116] (63% of the number of iOS devices)
Q1 2011	35%			393,000		Canalys, 4 May 2011.[117]
10 May 2011			100 million	400,000		Google I/O[118]
28 June 2011				500,000		a 4.4% weekly growth, Andy Rubin[119]
Q2 2011	48%	52%[120]		568,000		Canalys, 1 August 2011 [121]
July 14, 2011				550,000		4.4% growth per week. Google[122]
Q3 2011	52.5%			658,000		Gartner[114]
October 13, 2011			190 million			Google [123]
November 16, 2011			200 million			during the Google Music announcement "These Go to Eleven"[124] 3.8 million Android Honeycomb Tablets have been sold.[125]
December 20, 2011			250 million	700,000		Andy Rubin, Google[126]
27 February 2012			300 million	850,000		250% yearly growth rate. Andy Rubin, Google[127]
Q1 2012	59%		331 million	934,000		85 millions in 91 days, Signals and Systems Telecom[16]
27 June 2012			400 million	1 million		Google[17]

Usage share

Usage share of the different versions as of August 2, 2012.[128] Most Android devices to date run still the older OS version 2.3.x Gingerbread that was released on December 6, 2010.

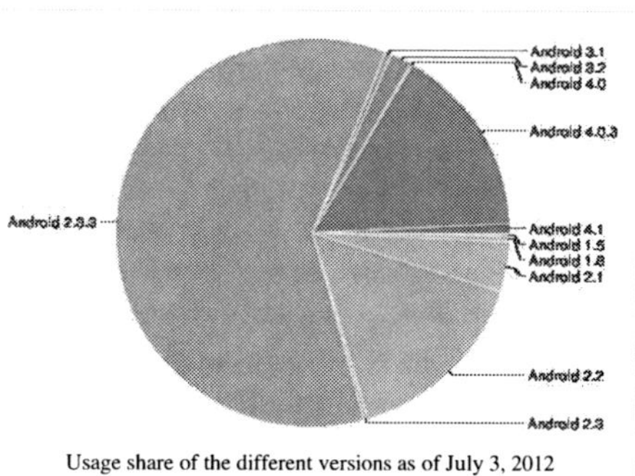

Usage share of the different versions as of July 3, 2012

Version	Release date	API level	Distribution (02 August 2012)
4.1.x *Jelly Bean*	July 10, 2012	16	0.8%
4.0.x *Ice Cream Sandwich*	October 19, 2011	14-15	15.9%
3.x.x *Honeycomb*	February 22, 2011	11-13	2.3%
2.3.x *Gingerbread*	December 6, 2010	9-10	60.6%
2.2 *Froyo*	May 20, 2010	8	15.5%
2.0, 2.1 *Eclair*	October 26, 2009	7	4.2%
1.6 *Donut*	September 15, 2009	4	0.5%
1.5 *Cupcake*	April 30, 2009	3	0.2%

Licensing

The source code for Android is available under free and open source software licenses. Google publishes most of the code (including network and telephony stacks)[129] under the Apache License version 2.0,[130][131][132] and the rest, Linux kernel changes, under the GNU General Public License version 2.

The Open Handset Alliance develops the changes to the Linux kernel, in public, with source code publicly available at all times. The rest of Android is developed in private, with source code released publicly when a new version is released. Typically Google collaborates with a hardware manufacturer to produce a flagship device (part of the Google Nexus series) featuring the new version of Android, then makes the source code available after that device has been released.[133]

In early 2011, Google chose to temporarily withhold the Android source code to the tablet-only Honeycomb release, the reason, according to Andy Rubin in an official Android blog post, was because Honeycomb was rushed for production of the Motorola Xoom,[134] and they did not want third parties creating a "really bad user experience" by attempting to put onto smartphones a version of Android intended for tablets.[135] The source code was once again made available in November 2011 with the release of Android 4.0.[136]

Copyrights and patents

Further information: Oracle v. Google

Both Android and Android phone manufacturers have been the target of numerous patent lawsuits. On August 12, 2010, Oracle sued Google over claimed infringement of copyrights and patents related to the Java programming language.[137] Oracle originally sought damages up to $6.1 billion,[138] but this valuation was rejected by a federal judge who asked Oracle to revise the estimate.[139] In response, Google submitted multiple lines of defense, counterclaiming that Android did not infringe on Oracle's patents or copyright, that Oracle's patents were invalid, and several other defenses. They said that Android is based on Apache Harmony, a clean room implementation of the Java class libraries, and an independently developed virtual machine called Dalvik.[140]

In May 2012 the jury in this case found that Google did not infringe on Oracle's patents, and the trial judge ruled that the structure of the Java APIs used by Google was not copyrightable.[141][142]

Microsoft has also sued several manufacturers of Android devices for patent infringement, and collects patent licensing fees from others. In October 2011 Microsoft said they had signed license agreements with ten Android device manufacturers, accounting for 55% of worldwide revenue for Android devices.[143] These include Samsung and HTC.[144]

Google has publicly expressed its dislike for the current patent landscape in the United States, accusing Apple, Oracle and Microsoft of trying to take down Android through patent litigation, rather than innovating and competing with better products and services.[145] In August 2011, Google started the process of purchasing Motorola Mobility for US$12.5 billion, which was viewed in part as a defensive measure to protect Android, since Motorola Mobility holds more than 17,000 patents.[146] In December 2011 Google bought over a thousand patents from IBM.[147]

References

[1] "Android Code Analysis" (http://www.ohloh.net/p/android/analyses/latest). . Retrieved 2012-06-01.

[2] "Philosophy and Goals" (http://source.android.com/about/philosophy.html). *Android Open Source Project*. Google. . Retrieved 2012-04-21.

[3] http://www.androidpolice.com/2012/06/27/breaking-jelly-bean-download-available-now-but-it-only-works-on-io-galaxy-nexuses-for-now/

[4] https://groups.google.com/forum/m/#!topic/android-building/XBYeD-bhk1o

[5] "MIPS gets sweet with Honeycomb" (http://www.eetimes.com/electronics-news/4215490/MIPS-gets-sweet-with-Honeycomb). Eetimes.com. . Retrieved 2012-02-20.

[6] Shah, Agam (December 1, 2011). "Google's Android 4.0 ported to x86 processors" (http://www.computerworld.com/s/article/9222323/Google_s_Android_4.0_ported_to_x86_processors). *Computerworld*. International Data Group. . Retrieved 2012-02-20.

[7] "Licenses" (http://source.android.com/source/licenses.html). *Android Open Source Project*. Google. . Retrieved 2012-02-20.

[8] http://www.android.com/

[9] Elgin, Ben (August 17, 2005). "Google Buys Android for Its Mobile Arsenal" (http://www.webcitation.org/5wk7sIvVb). *Bloomberg Businessweek*. Bloomberg. Archived from the original (http://www.businessweek.com/technology/content/aug2005/tc20050817_0949_tc024.htm) on February 24, 2011. . Retrieved 2012-02-20. "In what could be a key move in its nascent wireless strategy, Google (GOOG) has quietly acquired startup Android Inc. ..."

[10] "Industry Leaders Announce Open Platform for Mobile Devices" (http://www.openhandsetalliance.com/press_110507.html) (Press release). Open Handset Alliance. November 5, 2007. . Retrieved 2012-02-17.

[11] "Android Overview" (http://www.openhandsetalliance.com/android_overview.html). Open Handset Alliance. . Retrieved 2012-02-15.

[12] "About the Android Open Source Project" (http://source.android.com/about/index.html). . Retrieved 2012-02-20.

[13] Shankland, Stephen (November 12, 2007). "Google's Android parts ways with Java industry group" (http://news.cnet.com/8301-13580_3-9815495-39.html). *CNET News*. . Retrieved 2012-02-15.

[14] "Google Play hits 600,000 apps, 20 billion total installs" (http://www.engadget.com/2012/06/27/google-play-hits-600000-apps/?icid=eng_latest_art). .

[15] "Google's Android becomes the world's leading smart phone platform" (http://www.canalys.com/newsroom/googleâs-android-becomes-worldâs-leading-smart-phone-platform). *Canalys*. January 31, 2011. . Retrieved 2012-02-15.

[16] "Android Smartphone Activations Reached 331 Million in Q1'2012" (http://www.prweb.com/releases/2012/5/prweb9514037.htm). Signals and Systems Telecom. 2012-05-16. .

[17] "There have now been over 400 million Android devices activated, with over a million new activations every day" (https://plus.google.com/108967384991768947849/posts/jHLD6HTfx9U). official Android Engineering teams. 2012-06-27. .

[18] Greg Sandoval (August 2, 2010). "More signs iPhone under Android attack" (http://news.cnet.com/8301-13579_3-20012418-37.html). . Retrieved 2012-02-20.

[19] Markoff, John (November 4, 2007). "I, Robot: The Man Behind the Google Phone" (http://www.nytimes.com/2007/11/04/technology/04google.html?_r=2&hp=&pagewanted=all). *The New York Times*. . Retrieved 2012-02-15.

[20] Kirsner, Scott (September 2, 2007). "Introducing the Google Phone" (http://web.archive.org/web/20100104054533/http://www.boston.com/business/technology/articles/2007/09/02/introducing_the_google_phone/). *The Boston Globe*. Archived from the original (http://www.boston.com/business/technology/articles/2007/09/02/introducing_the_google_phone/) on January 4, 2010. . Retrieved 2012-02-15.

[21] Vogelstein, Fred (April 2011). "How the Android Ecosystem Threatens the iPhone" (http://www.wired.com/magazine/2011/04/mf_android/all/1). *Wired*. . Retrieved June 2, 2012.

[22] "T-Mobile Brings Unlimited Multiplayer Gaming to US Market with First Launch of Nokia N-Gage Game Deck" (http://newsroom.t-mobile.com/articles/t-mobile-nokia-N-Gage) (Press release). T-Mobile. September 23, 2003. . Retrieved 2012-02-15.

[23] Vance, Ashlee (August 7, 2011). "A Thousand Times Yes" (http://www.airsla.org/broadcasts/BusinessWeek110802.mp3). Bloomberg BusinessWeek. . Retrieved 2011-11-09.

[24] Block, Ryan (August 28, 2007). "Google is working on a mobile OS, and it's due out shortly" (http://www.engadget.com/2007/08/28/google-is-working-on-a-mobile-os-and-its-due-out-shortly). *Engadget*. . Retrieved 2012-02-17.

[25] Sharma, Amol; Delaney, Kevin J. (August 2, 2007). "Google Pushes Tailored Phones To Win Lucrative Ad Market" (http://online.wsj.com/article_email/SB118602176520985718-lMyQjAxMDE3ODA2MjAwMjIxWj.html). *The Wall Street Journal*. . Retrieved 2012-02-17.

[26] "Google admits to mobile phone plan" (http://web.archive.org/web/20070703031543/http://www.directtraffic.org/OnlineNews/Google_admits_to_mobile_phone_plan_18094880.html). *directtraffic.org*. Google News. March 20, 2007. Archived from the original (http://www.directtraffic.org/OnlineNews/Google_admits_to_mobile_phone_plan_18094880.html) on July 3, 2007. . Retrieved 2012-02-17.

[27] McKay, Martha (December 21, 2006). "Can iPhone become your phone?; Linksys introduces versatile line for cordless service" (http://record-bergen.vlex.com/vid/iphone-phone-linksys-versatile-cordless-62885923). *The Record (Bergen County)*: p. L9. . Retrieved 2012-02-21. "And don't hold your breath, but the same cell phone-obsessed tech watchers say it won't be long before Google jumps headfirst into the phone biz. Phone, anyone?"

[28] Claburn, Thomas (September 19, 2007). "Google's Secret Patent Portfolio Predicts gPhone" (http://www.informationweek.com/news/showArticle.jhtml?articleID=201807587&cid=nl_IWK_daily). *InformationWeek*. . Retrieved 2012-02-17.

[29] Pearce, James Quintana (September 20, 2007). "Google's Strong Mobile-Related Patent Portfolio" (http://www.moconews.net/entry/419-googles-strong-mobile-related-patent-portfolio). *mocoNews.net*. . Retrieved 2012-02-17.

[30] Martinez, Jennifer (December 10, 2008). "Corrected: Update 2: More mobile phone makers back Google's Android" (http://www.reuters.com/article/2008/12/10/openhandset-idUSN0928595620081210). Reuters. Thomson Reuters. . Retrieved 2012-02-16.

[31] Kharif, Olga (December 9, 2008). "Google's Android Gains More Powerful Followers" (http://www.businessweek.com/the_thread/techbeat/archives/2008/12/googles_android_2.html). *BusinessWeek*. . Retrieved 2012-02-16.

[32] http://source.android.com/

[33] "About the Android Open Source Project | Android Open Source" (http://source.android.com/about/index.html). Source.android.com. . Retrieved 2012-02-20.

[34] "Philosophy and Goals | Android Open Source" (http://source.android.com/about/philosophy.html). Source.android.com. . Retrieved 2012-02-20.

[35] "Frequently Asked Questions | Android Open Source" (http://source.android.com/faqs.html#compatibility). Source.android.com. . Retrieved 2012-02-20.

[36] John D. Sutter (February 4, 2011). "Why does Google name its Android products after desserts?" (http://articles.cnn.com/2011-02-04/tech/google.honeycomb.android.names_1_google-android-android-os-randall-sarafa?_s=PM:TECH). *CNNTech*. . Retrieved 2012-02-15.

[37] Tim Bray (November 24, 2010). "What Android Is" (http://www.tbray.org/ongoing/When/201x/2010/11/14/What-Android-Is). *ongoing by Tim Bray*. . Retrieved 2012-02-15.

[38] *Androidology – Part 1 of 3 – Architecture Overview* (http://www.youtube.com/watch?v=QBGfUs9mQYY) (Video). YouTube. September 6, 2008. . Retrieved 2007-11-07.

[39] Paul, Ryan (February 23, 2009). "Dream(sheep++): A developer's introduction to Google Android" (http://arstechnica.com/open-source/reviews/2009/02/an-introduction-to-google-android-for-developers.ars). *Ars Technica*. . Retrieved 2012-02-15.

[40] David Meyer (February 3, 2010). "Linux developer explains Android kernel code removal" (http://www.zdnet.com/news/linux-developer-explains-android-kernel-code-removal/389733). ZDNet. . Retrieved 2012-02-20.

[41] Greg Kroah-Hartman (February 2, 2010). "Android and the Linux kernel community" (http://www.kroah.com/log/linux/android-kernel-problems.html). . Retrieved 2012-02-20. "*Google shows no sign of working to get their code upstream anymore. Some companies are trying to strip the Android-specific interfaces from their codebase and push that upstream, but that causes a much larger engineering effort, and is a pain that just should not be necessary.*"

[42] Brian Proffitt (August 10, 2010). "Garrett's LinuxCon Talk Emphasizes Lessons Learned from Android/Kernel Saga" (https://www.linux.com/news/embedded-mobile/mobile-linux/344486-garretta-linuxcon-talk-emphasizes-lessons-learned-from-androidkernel-saga). Linux.com. . Retrieved 2012-02-21.

[43] Brian Proffitt (April 15, 2010). "DiBona: Google will hire two Android coders to work with kernel.org" (http://www.zdnet.com/blog/open-source/dibona-google-will-hire-two-android-coders-to-work-with-kernelorg/6274). *www.zdnet.com*. . Retrieved 2012-02-20.

[44] Steven J. Vaughan-Nichols (September 7, 2010). "Android/Linux kernel fight continues" (http://blogs.computerworld.com/16900/android_linux_kernel_fight_continues). Computerworld. . Retrieved 2012-02-20.

[45] Rafael J. Wysocki (November 24, 2010). "An alternative to suspend blockers" (http://lwn.net/Articles/416690/). *lwn.net*. . Retrieved 2012-02-15.

[46] Rafael J. Wysocki (November 12, 2010). "Technical Background of the Android Suspend Blockers Controversy" (http://lwn.net/images/pdf/suspend_blockers.pdf). . Retrieved 2012-02-15. "...the most controversial parts of the Android's opportunistic suspend infrastructure are not really necessary and therefore they should not be included into the mainline kernel.... it should be possible to convert the vast majority of the Android device drivers using wakelocks to the mainline kernel code base."

[47] Steven J. Vaughan-Nichols (August 18, 2011). "Linus Torvalds on Android, the Linux fork" (http://www.zdnet.com/blog/open-source/linus-torvalds-on-android-the-linux-fork/9426). *zdnet.com*. . Retrieved 2012-02-15.

[48] Chris von Eitzen (December 23, 2011). "Android drivers to be included in Linux 3.3 kernel" (http://www.h-online.com/open/news/item/Android-drivers-to-be-included-in-Linux-3-3-kernel-1400996.html). *h-online.com*. . Retrieved 2012-02-15.

[49] Swapnil Bhartiya (January 2, 2012). "Linux 3.3 Will Let You Boot Into Android: Greg-KH" (http://www.muktware.com/news/3273/linux-33-will-let-you-boot-android-greg-kh). *muktware.com*. . Retrieved 2012-02-15.

[50] "What is Android?" (http://developer.android.com/guide/basics/what-is-android.html). *Android Developers*. July 21, 2009. . Retrieved 2012-02-15.

[51] Topolsky, Joshua (November 12, 2007). "Google's Android OS early look SDK now available" (http://www.engadget.com/2007/11/12/googles-android-os-early-look-sdk-now-available/). *Engadget*. . Retrieved 2012-02-17.

[52] "Android Supported Media Formats" (http://developer.android.com/guide/appendix/media-formats.html). *Android Developers*. . Retrieved 2012-02-17.

[53] "Android 2.3 Platform Highlights" (http://developer.android.com/sdk/android-2.3-highlights.html). *Android Developers*. December 6, 2010. . Retrieved 2012-02-20.

[54] "Flash Flayer 10.1 for Android 2.2 Release Notes" (http://kb2.adobe.com/cps/860/cpsid_86018.html). *Adobe Knowledgebase*. . Retrieved 2012-02-16.

[55] "RealNetworks Gives Handset and Tablet OEMs Ability to Deliver HTTP Live Content to Android Users" (http://www.realnetworks.com/press/releases/2010/RealPlayer-for-Mobile-Delivers-HTTP-Live-Content-to-Android.aspx). *realnetworks.com*. September 10, 2010. . Retrieved 2012-02-16.

[56] "Android 3.0 Platform Highlights" (http://developer.android.com/sdk/android-3.0-highlights.html). *Google*. . Retrieved 2012-02-15.

[57] Musil, Steven (February 11, 2009). "Report: Apple nixed Android's multitouch" (http://news.cnet.com/8301-13579_3-10161312-37.html). *CNET News*. . Retrieved 2012-02-16.

[58] Ziegler, Chris (February 2, 2010). "Nexus One gets a software update, enables multitouch" (http://www.engadget.com/2010/02/02/nexus-one-gets-a-software-update-enables-multitouch/). *Engadget*. . Retrieved 2012-02-16.

[59] "Android 3.1 Platform Highlights" (http://developer.android.com/sdk/android-3.1-highlights.html#UserFeatures). *Android Developers*. . Retrieved 2012-02-16.

[60] Bray, Tim (April 28, 2010). "Multitasking the Android Way" (http://android-developers.blogspot.com/2010/04/multitasking-android-way.html). *Android Developers*. . Retrieved 2012-02-16.

[61] "Speech Input for Google Search" (http://developer.android.com/resources/articles/speech-input.html). *Android Developers*. . Retrieved 2012-02-16.

[62] "Voice Actions for Android" (http://www.google.com/mobile/voice-actions/). *google.com*. . Retrieved 2012-02-16.

[63] JR Raphael (May 6, 2010). "Use Your Android Phone as a Wireless Modem" (http://www.pcworld.com/article/190265/use_your_android_phone_as_a_wireless_modem.html). PCWorld. . Retrieved 2012-02-16.

[64] Nancy Gohring (October 19, 2011). "Samsung, Google Unveil Latest Android OS, Phone" (http://www.pcworld.com/article/242128/samsung_google_unveil_latest_android_os_phone.html). PCWorld. . Retrieved 2012-02-16.

[65] Laura June (September 6, 2010). "Toshiba AC100 Android smartbook hits the United Kingdom" (http://www.engadget.com/2010/09/06/toshiba-ac100-android-smartbook-hits-the-united-kingdom/). *Engadget*. . Retrieved 2012-02-20.

[66] Jolie O'Dell (May 12, 2011). "Androids Unite: How Ice Cream Sandwich Will End the OS Schism" (http://mashable.com/2011/05/12/ice-cream-sandwich/). *Mashable*. . Retrieved 2012-02-20.

[67] Hollister, Sean (January 10, 2012). "Sony Smart Watch (aka Sony Ericsson LiveView 2) hands-on" (http://www.theverge.com/2012/1/10/2695959/sony-smart-watch-aka-sony-ericsson-liveview-2-hands-on). The Verge. . Retrieved 2012-02-16.

[68] Rik Myslewski (January 12, 2011). "Android-powered touchscreen Wi-Fi headphones" (http://www.theregister.co.uk/2011/01/12/now_audio_admiral_touch/). theregister.co.uk. . Retrieved 2012-01-16.

[69] "Car Player Android-Car Player Android Manufacturers, Suppliers and Exporters on" (http://www.alibaba.com/showroom/car-player-android.html). Alibaba.com. . Retrieved 2012-02-20.

[70] Android Everywhere: 10 Types of Devices That Android Is Making Better (http://www.androidauthority.com/android-everywhere-10-types-of-devices-that-android-is-making-better-57012/)

[71] "Altek Leo, the 14 megapixel Android cameraphone, headed for Europe in 2011" (http://www.engadget.com/2010/10/03/altek-leo-the-14-megapixel-android-cameraphone-headed-for-euro/). Engadget. October 3, 2010. . Retrieved 2012-01-04.

[72] Will G. (December 1, 2011). "Top Android MP3 Players for 2011" (http://www.androidauthority.com/top-android-mp3-players-for-2011-36523/). Androidauthority.com. . Retrieved 2012-02-16.

[73] "Archos Smart Home Phone now available - get Android on your landline" (http://www.androidcentral.com/archos-smart-home-phone-now-available-get-android-your-landline). Android Central. January 19, 2012. . Retrieved 2012-01-30.

[74] ProForm Trailrunner 4.0 treadmill tricks you into exercising with 10-inch Android tablet - Engadget (http://www.engadget.com/2010/12/28/proform-trailrunner-4-0-treadmill-tricks-you-into-exercising-wit/)

[75] "T-Mobile Unveils the T-Mobile G1 - the First Phone Powered by Android" (http://web.archive.org/web/20110712230204/http://www.htc.com/www/press.aspx?id=66338&lang=1033). HTC. September 23, 2008. Archived from the original (http://www.htc.com/www/press.aspx?id=66338&lang=1033) on July 12, 2011. . Retrieved 2012-02-17. AT&T's first device to run the Android OS was the Motorola Backflip.

[76] Richard Wray (March 14, 2010). "Google forced to delay British launch of Nexus phone" (http://www.guardian.co.uk/technology/2010/mar/14/google-mobile-phone-launch-delay). London: guardian.co.uk. . Retrieved 2012-02-17.

[77] David Wang (May 19, 2010). "How to Install Android on Your iPhone" (http://www.pcworld.com/article/196595/how_to_install_android_on_your_iphone.html). pcworld.com. . Retrieved 2012-02-20.

[78] "iDroid Project Wiki" (http://www.idroidproject.org/). Idroidproject.org. . Retrieved 2012-02-20.

[79] Graziano, Dan (December 28, 2011). "Pentagon approves Android device for Department of Defense, Apple still awaits clearance" (http://www.bgr.com/2011/12/28/pentagon-approves-android-device-for-department-of-defense-apple-still-awaits-clearance/). Bgr.com. . Retrieved 2012-02-16.

[80] Ravi Mandalia (December 26, 2011). "Pentagon OKs Android for DoD Usage" (http://www.itproportal.com/2011/12/26/pentagon-oks-android-dod-usage/). ITProPortal.com. . Retrieved 2012-02-16.

[81] Chris Carroll (December 9, 2011). "Android to be Approved for DoD Use Within Weeks" (http://www.military.com/news/article/2011/android-to-be-approved-for-dod-use-within-weeks.html). Military.com. . Retrieved 2012-02-16.

[82] Ganapati, Priya (June 11, 2010). "Independent App Stores Take On Google's Android Market" (http://www.wired.com/gadgetlab/2010/06/independent-app-stores-take-on-googles-android-market/). Wired News. . Retrieved 2012-02-20.

[83] "Android Compatibility" (http://source.android.com/compatibility/index.html). *Android Open Source Project*. . Retrieved 2012-02-20.

[84] "Android Compatibility" (http://developer.android.com/guide/practices/compatibility.html). *Android Developers*. . Retrieved 2012-02-20.

[85] "Voice Actions for Android" (http://www.google.com/mobile/voice-actions/index.html). Google.com. . Retrieved 2012-02-20.

[86] "Android Security Overview" (http://source.android.com/tech/security/index.html). *Android Open Source Project*. . Retrieved 2012-02-20.

[87] Felt, Adrienne Porte; Chin, Erika; Hanna, Steve; Song, Dawn; Wagner, David. *Android Permissions Demystified* (http://www.cs.berkeley.edu/~afelt/android_permissions.pdf). . Retrieved 2012-02-20.

[88] Chin, Erika; Felt, Adrienne Porter; Greenwood, Kate; Wagner, David (2011). *Analyzing Inter-Application Communication in Android*. Mobisys.

[89] "Lookout Mobile Security" (http://www.mylookout.com). Lookout. . Retrieved 2012-07-05.

[90] "Antivirus for Android Smartphones" (http://www.avg.com/us-en/antivirus-for-android). AVG. . Retrieved 2012-02-16.

[91] "Mobile Security" (http://www.avast.com/free-mobile-security). Avast.com. . Retrieved 2012-02-16.

[92] "Mobile Security — System requirements" (http://www.f-secure.com/en/web/home_global/protection/mobile-security/system-requirements). F-Secure. . Retrieved 2012-02-16.

[93] "Kaspersky Mobile Security" (http://www.kaspersky.com/mobile_downloads). Kaspersky.com. . Retrieved 2012-02-16.

[94] "McAfee Mobile Security for Android" (https://www.mcafeemobilesecurity.com/products/android.aspx). Mcafeemobilesecurity.com. . Retrieved 2012-02-16.

[95] "Mobile Internet Security" (http://us.norton.com/mobile-security/). Us.norton.com. . Retrieved 2012-02-16.

[96] http://www.extremetech.com/computing/104827-android-antivirus-apps-are-useless-heres-what-to-do-instead/2 Android antivirus apps are useless, here's what to do instead — access April 10, 2012

[97] ReadWriteWeb: How to Find Your Lost or Stolen Android Phone for Free (http://www.readwriteweb.com/archives/how_to_find_a_lost_or_stolen_android_phone_for_free.php). Comparison of several "Find my Phone" applications for Android phones.

[98] Steve Lohr (May 8, 2011). "Suit Opens a Window Into Google" (http://www.nytimes.com/2011/05/09/technology/09google.html?scp=1&sq=Skyhook Wireless v.Google Case Yields E-Mail Insight&st=cse). *The New York Times*. ISSN 0362-4331. . Retrieved 2012-02-16.

[99] "AppAnalysis.org: Real Time Privacy Monitoring on Smartphones" (http://appanalysis.org/faq.html). . Retrieved 2012-02-21.

[100] Ganapati, Priya (September 30, 2010). "Study Shows Some Android Apps Leak User Data Without Clear Notifications | Gadget Lab" (http://www.wired.com/gadgetlab/2010/09/data-collection-android/). Wired.com. . Retrieved 2012-01-30.

[101] Et Tu, Google? Android Apps Can Also Secretly Copy Photos (http://bits.blogs.nytimes.com/2012/03/01/android-photos/), Bits blog from the New York Times, Retrieved March 2, 2012.

[102] Android allows apps to see your photos, like every computer does [FUD] | Android Central (http://www.androidcentral.com/android-allows-apps-see-your-photos-every-computer-does)

[103] Woyke, Elizabeth (September 26, 2008). "Android's Very Own Font" (http://www.forbes.com/2008/09/25/font-android-g1-tech-wire-cx_ew_0926font.html). *Forbes*. . Retrieved 2012-02-16.

[104] Murphy, Mark (2010) (in French), *L'art du développement Android 2* (http://books.google.de/books?hl=de&lr=&id=IDs63og2WpgC&oi=fnd&pg=PR13&dq=Android+"Irina+Blok"&ots=679uS18Z35&sig=GQFrExSOil1E1nVRPnPEmBjgo1A#v=onepage&q=Android "Irina Blok"&f=false), Pearson Education France, p. ii, ISBN 978-2-7440-2451-1,

[105] "Brand Guidelines" (http://www.android.com/branding.html). *Android*. . Retrieved 2012-02-16.

[106] "Smart phones defy slowdown" (http://www.canalys.com/newsroom/smart-phones-defy-slowdown). Canalys. 17 August 2009. .

[107] "Smart phone market shows modest growth in Q3" (http://www.canalys.com/newsroom/smart-phone-market-shows-modest-growth-q3). Canalys. 3 November 2009. .

[108] NPD group, May 2010 "Android hits top spot in U.S. smartphone market" (http://news.cnet.com/8301-1035_3-20012627-94.html). August 4, 2010. . Retrieved 2012-02-20.

[109] "North American smart phone shipments to exceed 65 million units in 2010" (http://www.canalys.com/newsroom/north-american-smart-phone-shipments-exceed-65-million-units-2010). Canalys. 17 March 2010. .

[110] "comScore Reports February 2010 U.S. Mobile Subscriber Market Share". Comscore. April 5, 2010.

[111] David Sarno (12 May 2010). [articles.latimes.com/2010/may/12/business/la-fi-google-apple-20100512 "Google's Android smart-phone sales leapfrog Apple's iPhone"]. Los Angeles Times. articles.latimes.com/2010/may/12/business/la-fi-google-apple-20100512.

[112] Charles Arthur (23 June 2010). "Google 'activating 160,000 Android phones a day'" (http://www.guardian.co.uk/technology/2010/jun/23/google-android-eric-schmidt). The Guardian. .

[113] "Apple takes the lead in the US smart phone market with a 26% share" (http://www.canalys.com/newsroom/apple-takes-lead-us-smart-phone-market-26-share). Canalys. 1 november 2010. .

[114] "Gartner Says Sales of Mobile Devices Grew 5.6 Percent in Third Quarter of 2011; Smartphone Sales Increased 42 Percent" (http://www.gartner.com/it/page.jsp?id=1848514). November 15, 2011. . Retrieved 2012-02-16.

[115] "comScore Reports September 2010 U.S. Mobile Subscriber Market Share" (http://www.comscore.com/Press_Events/Press_Releases/2010/11/comScore_Reports_September_2010_U.S._Mobile_Subscriber_Market_Share). *Comscore.com*. November 3, 2010. . Retrieved

2012-02-20.
[116] "Apple iOS Platform Outreaches Android by 59 Percent in U.S. When Accounting for Mobile Phones, Tablets and Other Connected Media Devices" (http://www.comscore.com/Press_Events/Press_Releases/2011/4/Apple_iOS_Platform_Outreaches_Android_by_59_Percent_in_U.S). *comScore*. April 19, 2011. . Retrieved 2012-02-20.
[117] "Android increases smart phone market leadership with 35% share" (http://www.canalys.com/newsroom/android-takes-almost-50-share-worldwide-smart-phone-market). 4 May 2011. . Retrieved 2012-06-05.
[118] Barra, Hugo (May 10, 2011). "Android: momentum, mobile and more at Google I/O" (http://googleblog.blogspot.com/2011/05/android-momentum-mobile-and-more-at.html). *The Official Google Blog*. . Retrieved 2012-02-16.
[119] Andy Rubin (Jun 28, 2011). "" (https://twitter.com/arubin/statuses/85660213478309888). .
[120] "As Android Solidifies Lead, Google Acquisition Has Potential to Revitalize Flagging Motorola" (https://www.npd.com/wps/portal/npd/us/news/pressreleases/pr_110822a/). The NPD Group. 22 August 2011. .
[121] "Android takes almost 50% share of worldwide smart phone market" (http://www.canalys.com/newsroom/android-takes-almost-50-share-worldwide-smart-phone-market). August 1, 2011. . Retrieved 2012-02-16.
[122] Kumparak, Greg (July 14, 2011). "Android Now Seeing 550,000 Activations Per Day" (http://techcrunch.com/2011/07/14/android-now-seeing-550000-activations-per-day/). *Techcrunch*. . Retrieved 2012-02-16.
[123] Erick Schonfeld (October 13, 2011). "Larry Page: Mobile Revenues At $2.5 Billion Run-Rate, 190 Million Android Devices" (http://techcrunch.com/2011/10/13/page-google-plus-40-million-mobile-2-5-billion/). *TechCrunch*. . Retrieved 2012-02-16.
[124] Lance Whitney (November 17, 2011). "Google: 200 million Android devices now active worldwide" (http://news.cnet.com/8301-1023_3-57326649-93/google-200-million-android-devices-now-active-worldwide/). *CNET News*. . Retrieved 2012-02-16.
[125] Charlie Sorrel (October 19, 2011). "Only 3.8 Million Honeycomb Tablets Sold So Far" (http://www.wired.com/gadgetlab/2011/10/only-3-8-million-honeycomb-tablets-sold-so-far/). Wired.com. . Retrieved 2012-02-16.
[126] Schonfeld, Erick (December 22, 2011). "Android Phones Pass 700,000 Activations Per Day, Approaching 250 Million Total" (http://techcrunch.com/2011/12/22/android-700000/). *TechCrunch*. AOL. . Retrieved 2012-02-15.
[127] Andy Rubin (February 27, 2012). "Android@Mobile World Congress: It's all about the ecosystem." (http://googlemobile.blogspot.fr/2012/02/androidmobile-world-congress-its-all.html). .
[128] "Android Platform Versions" (http://developer.android.com/resources/dashboard/platform-versions.html). *Android Developers*. May 1, 2012. . Retrieved May 2, 2012. "Based on the number of Android devices that have accessed Android Market within a 14-day period ending on the data collection date noted below."
[129] Boulton, Clint (October 21, 2008). "Google Open-Sources Android on Eve of G1 Launch" (http://www.eweek.com/c/a/Mobile-and-Wireless/Google-Open-Sources-Android-on-Eve-of-G1-Launch/). *eWeek*. . Retrieved 2012-02-17.
[130] Bort, Dave (October 21, 2008). "Android is now available as open source" (http://web.archive.org/web/20090228170042/http://source.android.com/posts/opensource). *Android Open Source Project*. Archived from the original (http://source.android.com/posts/opensource) on February 28, 2009. . Retrieved 2012-02-16.
[131] "Licenses: Android Open Source" (http://source.android.com/source/licenses.html). *Android Open Source Project*. . Retrieved 2012-02-16.
[132] Ryan Paul (November 6, 2007). "Why Google chose the Apache Software License over GPLv2 for Android" (http://arstechnica.com/old/content/2007/11/why-google-chose-the-apache-software-license-over-gplv2.ars). *Ars Technica*. . Retrieved 2012-02-16.
[133] "Frequently Asked Questions: What is involved in releasing the source code for a new Android version?" (http://source.android.com/faqs.html#what-is-involved-in-releasing-the-source-code-for-a-new-android-version). *Android Open Source Project*. . Retrieved 2012-02-16.
[134] Bray, Tim (April 6, 2011). "Android Developers Blog: I think I'm having a Gene Amdahl moment" (http://android-developers.blogspot.com/2011/04/i-think-im-having-gene-amdahl-moment.html). Android-developers.blogspot.com. . Retrieved 2012-02-16.
[135] Jerry Hildenbrand (March 24, 2011). "Honeycomb won't be open-sourced? Say it ain't so!" (http://www.androidcentral.com/google-not-open-sourcing-honeycomb-says-bloomberg). Androidcentral.com. . Retrieved 2012-02-16.
[136] Thom Holwerda (November 14, 2011). "Android 4.0 Ice Cream Sandwich Source Code Released" (http://www.osnews.com/story/25330/Android_4_0_Ice_Cream_Sandwich_Source_Code_Released). *OSNews*. . Retrieved 2012-02-16.
[137] Niccolai, James (August 12, 2010). "Update: Oracle sues Google over Java use in Android" (http://www.computerworld.com/s/article/9180678/Update_Oracle_sues_Google_over_Java_use_in_Android). *Computerworld*. International Data Group Inc. . Retrieved 2012-02-16.
[138] "Oracle seeks up to $6.1 billion in Google lawsuit" (http://www.reuters.com/article/2011/06/18/us-oracle-google-lawsuit-idUSTRE75H0FP20110618). Reuters. June 18, 2011. . Retrieved September 7, 2011.
[139] "Judge tosses Oracle's $6.1 billion damage estimate in claim against Google" (http://www.mercurynews.com/news/ci_18532705). MercuryNews.com. July 22, 2011. . Retrieved September 7, 2011.
[140] Singel, Ryan (October 5, 2010). "Calling Oracle Hypocritical, Google Denies Patent Infringement" (http://www.wired.com/epicenter/2010/10/google-oracle-android/). *Wired*. . Retrieved 2012-02-16.
[141] Josh Lowensohn (May 23, 2012). "Jury clears Google of infringing on Oracle's patents" (http://www.zdnet.com/blog/btl/jury-clears-google-of-infringing-on-oracle-patents/77897). *ZDNet*. . Retrieved 2012-05-25.
[142] Joe Mullin (May 31, 2012). "Google wins crucial API ruling, Oracle's case decimated" (http://arstechnica.com/tech-policy/2012/05/google-wins-crucial-api-ruling-oracles-case-decimated/). *Ars Technica*. . Retrieved 2012-06-01.
[143] "Microsoft collects license fees on 50% of Android devices, tells Google to "wake up"" (http://arstechnica.com/microsoft/news/2011/10/microsoft-collects-license-fees-on-50-of-android-devices-tells-google-to-wake-up.ars). *Ars Technica*. . Retrieved 2012-02-16.

[144] Mikael Ricknäs (September 28, 2011). "Microsoft signs Android licensing deal with Samsung" (http://www.computerworld.com/s/article/9220357/Microsoft_signs_Android_licensing_deal_with_Samsung). *Computerworld*. . Retrieved 2012-02-16.

[145] Jacqui Cheng (August 3, 2011). "Google publicly accuses Apple, Microsoft, Oracle of patent bullying" (http://arstechnica.com/tech-policy/news/2011/08/google-publicly-accuses-apple-microsoft-oracle-of-patent-bullying.ars). . Retrieved 2012-02-16.

[146] Casey Johnston (August 15, 2011). "Google, needing patents, buys Motorola wireless for $12.5 billion" (http://arstechnica.com/gadgets/news/2011/08/google-to-buy-motorola-in-effort-to-defend-itself-from-patent-bullies.ars). . Retrieved 2012-02-16.

[147] Paul, Ryan (January 4, 2012). "Google buys another round of IBM patents as its Oracle trial nears" (http://arstechnica.com/gadgets/news/2012/01/google-buys-another-round-of-ibm-patents-as-oracle-trial-nears.ars). Ars Technica (http://arstechnica.com/). . Retrieved 2012-02-16.

External links

- Official website (http://www.android.com/)
- Android devices at Google.com (http://www.google.com/phone/)
- Google's Android apps (http://www.google.com/mobile/android/)
- Android (operating system) (http://www.dmoz.org/Computers/Systems/Handhelds/Android/) at the Open Directory Project
- Sergey Brin introduces the Android platform (https://www.youtube.com/watch?v=1FJHYqE0RDg) on YouTube
- Android: Building a Mobile Platform to Change the Industry (http://www.stanford.edu/class/ee380/Abstracts/071128.html): lecture given by Google Mobile Platforms Manager, Richard Miner at Stanford University (video archive (http://ee380.stanford.edu/cgi-bin/videologger.php?target=071128-ee380-300.asx))
- Android Internals: Fragment of a course detailing the architecture of Android and interaction of its components (http://technologeeks.com/Courses/Android-Excerpt.pdf)

Samsung

Samsung Group

□□□□

□□□□

Type	Chaebol
Industry	Conglomerate
Founded	1938
Founder(s)	Lee Byung-chull
Headquarters	Samsung Town, Seoul, South Korea
Area served	Worldwide
Key people	Lee Kun-hee (Chairman and CEO) Lee Soo-bin (President, CEO of Samsung Life Insurance)[1]
Products	Apparel, chemicals, consumer electronics, electronic components, medical equipment, precision instruments, semiconductors, ships, telecommunications equipment
Services	Advertising, construction, entertainment, financial services, hospitality, information and communications technology services, medical services, retail
Revenue	US$ 247.5 billion (2011)[2]
Net income	US$ 18.3 billion (2011)[2]
Total assets	US$ 384.3 billion (2011)[2]
Total equity	US$ 224.7 billion (2011)[2]
Employees	344,000 (2010)[2]
Subsidiaries	Samsung Electronics Samsung Life Insurance Samsung Heavy Industries Samsung C&T Samsung SDS etc.
Website	Samsung.com [3]

Samsung Group (Hangul: □□□□; Hanja: □□□□; Korean pronunciation: [sam.sʌŋ ɡɯ'ɾup̚]) is a South Korean multinational conglomerate company headquartered in Samsung Town, Seoul. It comprises numerous subsidiaries and affiliated businesses, most of them united under the *Samsung* brand, and is the largest South Korean *chaebol*.

Notable Samsung industrial subsidiaries include Samsung Electronics (the world's largest information technology company measured by 2011 revenues),[4][5] Samsung Heavy Industries (the world's second-largest shipbuilder measured by 2010 revenues),[6] Samsung Engineering and Samsung C&T (respectively the world's 35th- and 72nd-largest construction companies), and Samsung Techwin (a weapons technology and optoelectronics manufacturer).[7] Other notable subsidiaries include Samsung Life Insurance (the world's 14th-largest life insurance company),[8] Samsung Everland (the oldest theme park in South Korea)[9] and Cheil Worldwide (the world's 19th-largest advertising agency measured by 2010 revenues).[10][11]

Samsung produces around a fifth of South Korea's total exports[12] and its revenues are larger than many countries' GDP; in 2006, it would have been the world's 35th-largest economy.[13] The company has a powerful influence on South Korea's economic development, politics, media and culture, and has been a major driving force behind the "Miracle on the Han River".[14][15]

Name

According to the founder of Samsung Group, the meaning of the Korean hanja word *Samsung* (□ □) is "tristar" or "three stars". The word "three" represents something "big, numerous and powerful"; the "stars" mean eternity.[16]

History

1938 to 1970

In 1938,[17] Lee Byung-chull (1910–1987) of a large landowning family in the Uiryeong county came to the nearby Daegu city and founded *Samsung Sanghoe* (□ □ □ □), a small trading company with forty employees located in Su-dong (now Ingyo-dong). It dealt in groceries produced in and around the city and produced its own noodles. The company prospered and Lee moved its head office to Seoul in 1947. When the Korean War broke out, however, he was forced to leave Seoul and started a sugar refinery in Busan as a name of *Cheil Jedang*. After the war, in 1954, Lee founded *Cheil Mojik* and built the plant in Chimsan-dong, Daegu. It was the largest woollen mill ever in the country and the company took on an aspect of a major company.

The headquarters of *Samsung Sanghoe* in Daegu in the late-1930s

Samsung diversified into many areas and Lee sought to establish Samsung as an industry leader in a wide range of enterprises, moving into businesses such as insurance, securities, and retail. Lee placed great importance on industrialization, and focused his economic development strategy on a handful of large domestic conglomerates, protecting them from competition and assisting them financially.

In 1948, Cho Hong-jai (the Hyosung group's founder) jointly invested in a new company called Samsung Mulsan Gongsa (□ □ □ □ □ □), or the Samsung Trading Corporation, with the Samsung Group founder Lee Byung-chull. The trading firm grew to become the present-day Samsung C&T Corporation. But after some years Cho and Lee separated due differences in management between them. He wanted to get up to a 30% group share. After settlement, Samsung Group was separated into Samsung Group and Hyosung Group, Hankook Tire ...etc.[18][19]

In the late 1960s, Samsung Group entered into the electronics industry. It formed several electronics-related divisions, such as Samsung Electronics Devices Co., Samsung Electro-Mechanics Co., Samsung Corning Co., and Samsung Semiconductor & Telecommunications Co., and made the facility in Suwon. Its first product was a black-and-white television set.

1970 to 1990

The SPC-1000, introduced in 1982, was Samsung's first personal computer (Korean market only) and uses an audio cassette tape to load and save data - the floppy drive was optional[20]

In 1980, Samsung acquired the Gumi-based *Hanguk Jeonja Tongsin* and entered the telecommunications hardware industry. Its early products were switchboards. The facility were developed into the telephone and fax manufacturing systems and became the centre of Samsung's mobile phone manufacturing. They have produced over 800 million mobile phones to date.[21] The company grouped them together under Samsung Electronics Co., Ltd. in the 1980s.

After the founder's death in 1987, Samsung Group was separated into four business groups - Samsung Group, Shinsegae Group, CJ Group and Hansol Group.[22] Shinsegae (discount store, department store) was originally part of Samsung Group, separated in the 1990s from the Samsung Group along with CJ Group (Food/Chemicals/Entertainment/logistics) and the Hansol Group (Paper/Telecom). Today these separated groups are independent and they are not part of or connected to the Samsung Group.[23] One Hansol Group representative said, "Only people ignorant of the laws governing the business world could believe something so absurd," adding, "When Hansol separated from the Samsung Group in 1991, it severed all payment guarantees and share-holding ties with Samsung affiliates." One Hansol Group source asserted, "Hansol, Shinsegae, and CJ have been under independent management since their respective separations from the Samsung Group." One Shinsegae Department Store executive director said, "Shinsegae has no payment guarantees associated with the Samsung Group."[23]

In the 1980s, Samsung Electronics began to invest heavily in research and development, investments that were pivotal in pushing the company to the forefront of the global electronics industry. In 1982, it built a television assembly plant in Portugal; in 1984, a plant in New York; in 1985, a plant in Tokyo; in 1987, a facility in England; and another facility in Austin in 1996. In total, Samsung has invested about $5.6 billion in the Austin location – by far the largest foreign investment in Texas and one of the largest single foreign investments in the United States. The new investment will bring the total Samsung investment in Austin to more than $9 billion.[24]

1990 to 2000

Presamsung group;1938~1987

Hansol	Shinsegae	CJ	Samsung
Ownership; founder's eldest **daughter**	Ownership; founder's fifth **daughter**	Ownership; founder's eldest son	Ownership; founder's third son
1993 spun off	1997 spun off	1997 spun off	

Samsung started to rise as an international corporation in the 1990s. Samsung's construction branch was awarded a contract to build one of the two Petronas Towers in Malaysia, Taipei 101 in Taiwan and the Burj Khalifa in United Arab Emirates.[25] In 1993, Lee Kun-hee sold off ten of Samsung Group's subsidiaries, downsized the company, and merged other operations to concentrate on three industries: electronics, engineering, and chemicals. In 1996, the Samsung Group reacquired the Sungkyunkwan University foundation.

Samsung became the largest producer of memory chips in the world in 1992, and is the world's second-largest chipmaker after Intel (see Worldwide Top 20 Semiconductor Market Share Ranking Year by Year).[26] In 1995, it created its first liquid-crystal display screen. Ten years later, Samsung grew to be the world's largest manufacturer of liquid-crystal display panels. Sony, which had not invested in large-size TFT-LCDs, contacted Samsung to cooperate, and, in 2006, S-LCD was established as a joint venture between Samsung and Sony in order to provide a stable supply of LCD panels for both manufacturers. S-LCD was owned by Samsung (50% plus 1 share) and Sony (50% minus 1 share) and operates its factories and facilities in Tangjung, South Korea. As on 26 December 2011 it was announced that Samsung had acquired the stake of Sony in this joint venture.[27]

Compared to other major Korean companies, Samsung survived the 1997 Asian financial crisis relatively unharmed. However, Samsung Motor was sold to Renault at a significant loss. As of 2010, Renault Samsung is 80.1 percent owned by Renault and 19.9 percent owned by Samsung. Additionally, Samsung manufactured a range of aircraft from the 1980s to 1990s. The company was founded in 1999 as Korea Aerospace Industries (KAI), the result of merger between then three domestic major aerospace divisions of Samsung Aerospace, Daewoo Heavy Industries, and Hyundai Space and Aircraft Company. However, Samsung still manufactures aircraft engines and gas turbines.[28]

2000 to present

Samsung Techwin has been the sole supplier of a combustor module of the Trent 900 engine of the Rolls-Royce Airbus A380-The largest passenger airliner in the world- since 2001.[29] Samsung Techwin of Korea is a revenue-sharing participant in the Boeing's 787 Dreamliner GEnx engine program.[30]

The Samsung pavilion at Expo 2012

Samsung was instrumental in bringing the 2018 Winter Olympics to Pyeongchang. In December 2009, the former chairman of Samsung, Lee Kun-hee, was pardoned in order that he could return to the International Olympics Committee and help South Korea bid for the 2018 Winter Olympics in Pyeongchang. He had previously been convicted of tax evasion in 2008 and had been part of two failed bids to bring the Olympics to South Korea.[31] During the bid, Lee Kun-hee and figure skating gold medalist Kim Yuna lobbied heavily for support; it was thought that Lee's influence would help to secure the bid. On July 6, 2011, it was announced that Pyeongchang would be the location of the 2018 Winter Games.[32] Samsung C&T Corporation will be among the top tier of firms competing for construction projects for the games.[33]

In 2010, Samsung announced a 10-year growth strategy centered around five businesses.[34] One of these businesses was to be focused on biopharmaceuticals, to which the Company has committed ₩2.1 trillion.[35]

In December 2011, Samsung Electronics sold its hard disk drive (HDD) business to Seagate.[36]

In the first quarter of 2012, Samsung Electronics became the world's largest mobile phone maker by unit sales, overtaking Nokia, which had been the market leader since 1998.[37][38]

Acquisitions and attempted acquisitions

For a company of its size Samsung has made relatively few acquisitions.[39]

Rollei – Swiss watch battle

> Samsung Techwin acquired a German camera-maker Rollei on 1995. Samsung (Rollei) used its optic expertise on the crystals of a new line of 100% Swiss-made watches, designed by a team of watchmakers at Nouvelle Piquerez S.A. in Bassequort, Switzerland. Rolex's decision to fight Rollei on every front stemmed from the close resemblance between the two names and fears that its sales would suffer as a consequence. In the face of such a threat, the Geneva firm decided to confront. Rolex, this was also a demonstration of the Swiss watch industry's determination to defend itself when an established brand is threatened. Rolex sees this front-line

battle as vital for the entire Swiss watch industry. Rolex has succeeded in keeping Rollei out of the German market. On 11 March 1995 the Cologne District court prohibited the advertising and sale of Rollei watches on German territory.[40][41]

Fokker, a Dutch aircraft maker

Samsung lost a chance to revive its failed bid to take over Dutch aircraft maker Fokker when other airplane makers rejected its offer to form a consortium. The three proposed partners – Hyundai, Hanjin and Daewoo – have notified the South Korean government that they will not join Samsung Aerospace Industries Ltd.[42]

AST Research

Samsung bought AST (1994) and tried to break into North America, but the effort foundered. Samsung was forced to close the California-based computer maker following mass defection of research staff and a string of losses.[43]

FUBU clothing and apparel

In 1992, Daymond John had started the company with a hat collection that was made in his house in the Queens area of New York City. To fund the company, John had to mortgage his house for $100,000. With his friends, namely J. Alexander Martin, Carl Brown and Keith Perrin, half of his house was turned into the first factory of FUBU, while the other half remained as the living quarters. Along with the expansion of FUBU, Samsung, a Korean company, invested in FUBU in 1995.[44]

Lehman Brothers Holdings' Asian operations

Samsung Securities was one of a handful of brokerages looking into Lehman Brothers Holdings. But Nomura Holdings has reportedly waved the biggest check to win its bid for Lehman Brothers Holdings' Asian operations, beating out Samsung Securities, Standard Chartered, and Barclays.[45] Ironically, after few months Samsung Securities Co., Ltd. and City of London-based N M Rothschild & Sons (more commonly known simply as **Rothschild**) have agreed to form a strategic alliance in investment banking business. Two parties will jointly work on cross border mergers and acquisition deals.[46]

Grandis Inc. - memory developer

In July 2011, Samsung announced that it had acquired spin-transfer torque random access memory (MRAM) vendor Grandis Inc.[47] Grandis will become a part of Samsung's R&D operations and will focus on development of next generation random-access memory.[48]

Samsung and Sony joint venture - LCD display

On December 26, 2011 the board of Samsung Electronics approved a plan to buy Sony's entire stake in their 2004 joint liquid crystal display (LCD) venture for 1.08 trillion won ($938.97 million).[49]

Operations

Samsung Group headquarters at Samsung Town, Seoul

Samsung comprises around 80 companies.[50] It is highly diversified, with activities in areas including construction, electronics, financial services, shipbuilding and medical services.[50]

In FY 2009, Samsung reported consolidated revenues of 220 trillion KRW ($172.5 billion). In FY 2010, Samsung reported consolidated revenues of 280 trillion KRW ($258 billion), and profits of 30 trillion KRW ($27.6 billion) (based upon a KRW-USD exchange rate of 1,084.5 KRW per USD, the spot rate as of 19 August 2011).[51] However it should be noted that these amounts do not include the revenues of all subsidiaries of the group which are based outside South Korea.[52]

Subsidiaries and affiliates

As of April 2011 the Samsung Group comprised 59 unlisted companies and 19 listed companies, all of which had their primary listing on the Korea Exchange stock-exchange.[53]

The headquarters of Samsung Engineering in Seoul

Company	Symbol	Company	Symbol
Samsung C&T Corporation	000830	Shilla Hotels and Resorts	008770
Samsung Securities	016360	Samsung Fine Chemicals	004000
Samsung SDI	006400	SI Corporation	012750
Samsung Electro-Mechanics	009150	Samsung Fire & Marine Insurance	000810
Samsung Engineering	028050	Samsung Electronics	005930
Samsung Techwin	012450	Samsung Life Insurance	032830
Cheil Industries	001300	Samsung Card	029780
Samsung Heavy Industries	010140	Cheil Worldwide	030000
Imarket Korea	122900	Credu	067280
Ace Digitech	036550		

Samsung Electro-Mechanics

Samsung Electro-Mechanics, established in 1973 as a manufacturer of key electronic components, is headquartered in Suwon, Kyeonggi-Do, South Korea.

Samsung Machine Tools

Samsung Machine Tools of America is a national distributor of machines in the United States.[54]

Joint ventures

Current

- **aT Grain**

State-run Korea Agro-Fisheries Trade Corp. set up the venture, aT Grain Co., in Chicago, with three other South Korean companies, Korea Agro-Fisheries owns 55 percent of aT Grain, while Samsung C&T Corp, Hanjin Transportation Co. and STX Corporation. each hold 15 percent.[55]

- **Brooks Automation Asia**

Brooks Automation Asia Co., Ltd. is a joint venture between Brooks Automation (70%) and Samsung (30%) which was established in 1999. The venture locally manufactures and configure vacuum wafer handling platforms and 300mm Front-Opening Unified Pod (FOUP) load port modules, and designs, manufactures and configures atmospheric loading systems for flat panel displays.[56]

- **POSCO-SAMSUNG Slovakia Steel Processing Center Co., Ltd.** (Slovakia)

Company POSS - SLPC s.r.o. was founded in 2007 as a subsidiary of Samsung C & T Corporation, Samsung C & T Deutschland and the company POSCO.[57]

- **POSCO SAMSUNG Suzhou Steel Processing Center Co., Ltd.** (China)[58]
- **Samsung Air China Life Insurance**

Samsung Air China Life Insurance Co., Ltd. is a 50:50 joint venture between Samsung Life Insurance and China National Aviation Corporation. It was established in Beijing in July 2005.[59]

- **Samsung Bioepis**

Samsung Bioepis is a joint venture between Samsung Biologics (85%) and the United States-based Biogen Idec (15%).[60]

- **Samsung Biologics**

Samsung Electronics Co. and Samsung Everland Inc. will each own a 40 percent stake in the venture, with Samsung C&T Corp. and Durham, North Carolina-based Quintiles each holding 10 percent. It will contract-make medicines made from living cells, and Samsung Group plans to expand into producing copies of biologics including Rituxan, the leukemia and lymphoma treatment sold by Roche Holding AG and Biogen Idec Inc.[61]

- **Samsung BP Chemicals**

Samsung BP Chemicals Co., Ltd is a 50:50 joint venture between Samsung and the United Kingdom-based BP, which was established in 1989 to produce and supply high-value-added chemical products.

- **Samsung Corning Precision Glass**

Samsung Corning Precision Glass is a joint venture between Samsung and Corning, which was established in 1973 to manufacture and market cathode ray tube glass for black and white televisions. The company's first LCD glass substrate manufacturing facility opened in Gumi, Korea in 1996.

- **Samsung Sumitomo LED Materials**

Samsung Sumitomo LED Materials is a Korea-based joint venture between Samsung LED Co., Ltd., an LED maker based in Suwon, Korea-based and the Japan-based Sumitomo Chemical. The JV will carry out research and development, manufacturing, and sales of sapphire substrates for LEDs.[62]

- **Samsung Thales**

Samsung Thales Co., Ltd. (until 2001 known as Samsung Thomson-CSF Co., Ltd.) is a joint venture between Samsung Techwin and the France-based aerospace and defence company Thales. It was establsihed in 1978 and is based in Seoul.[63]

- **Samsung Total**

Samsung Total is a 50:50 joint venture between Samsung and the France-based oil group Total S.A. (more specifically Samsung General Chemicals and Total Petrochemicals).

- **SB LiMotive**

SB LiMotive is a 50:50 joint company of Robert Bosch GmbH(commonly known as Bosch) and Samsung SDI founded in June 2008. The joint venture develops and manufactures lithium-ion batteries for use in hybrid-, plug-in hybrid vehicles and electric vehicles.

- **SD Flex**

SD Flex Co., Ltd. was founded on October 2004 as a joint venture corporation by Samsung and DuPont, one of the world's largest chemical company.[64]

- **Sermatech Korea** (1999 to present)

Sermatech owns 51% of its stock, while Samsung owns the remaining 49%. The U.S. firm Sermatech International, for a business focusing on highly specialized aircraft construction processes such as special welding and brazing.[65]

- **Siam Samsung Life Insurance**

Samsung Life Insurance, holds a 37% stake while Saha Group also has a 37.5% stake in the joint venture, with the remaining 25% owned by Thanachart Bank.[66]

- **Siltronic Samsung Wafer**

Siltronic Samsung Wafer Pte. Ltd, the joint venture by Samsung and wholly owned Wacker Chemie subsidiary Siltronic, was officially opened in Singapore in June 2008.[67]

- **SMP**

SMP Ltd. is a joint venture between Samsung Fine Chemicals and MEMC. MEMC Electronic Materials Inc. and an affiliate of Korean conglomerate Samsung are forming a joint venture to build a polysilicon plant.

- **Steco**

Steco Co., Ltd. is established as the joint - venture company with Samsung Electronics Co., Ltd and Japan TORAY in 1995.[68]

- **Stemco**

Stemco is a joint venture between Samsung Electro-Mechanics and the Japan-based Toray Industries which was established in 1995.[69]

- **Toshiba Samsung Storage Technology**

Toshiba Samsung Storage Technology Corporation (TSST) is joint venture between Samsung Electronics and Toshiba of Japan which specialises in optical disc drive manufacturing. TSST was formed in 2004, and Toshiba owns 51% of its stock, with Samsung owns the remaining 49%.

Defunct

- **Alpha Processor**

In 1998, Samsung created a U.S. joint venture with Compaq--called Alpha Processor Inc. (API)--to help it enter the high-end processor market. The venture was also aimed at expanding Samsung's non-memory chip business. At the time, Samsung and Compaq invested $500 million in Alpha Processor.[70]

- **GE-Samsung Lighting**

GE Samsung Lighting was a joint venture between Samsung and the GE Lighting subsidiary of General Electric. The venture was established in 1998 and was broken up 2009. [71]

- **Global Steel Exchange**

Global Steel Exchange was a joint venture formed in 2000 between Samsung, the United States-based Cargill, the Switzerland-based Duferco Group, and the Luxembourg-based Tradearbed (now part of the ArcelorMittal), to handle their online buying and selling of steel.[72]

- **S-LCD**

S-LCD Corporation was a joint venture between Samsung Electronics (50% plus one share) and the Japan-based Sony Corporation (50% minus one share) which was established in April 2004. On December 26, 2011, Samsung Electronics announced that it would acquire all of Sony's shares in the venture.

Partially owned companies

- **Atlantico Sul**

Samsung Heavy Industries currently owns 10 percent of the Brazilian shipbuilder Atlantico Sul, which operates the largest shipyard in South America. Joao Candido, the largest ship built to date in Brazil, was built by Atlantico Sul with technology supplied by Samsung Heavy Industries.[73]

- **DGB Financial Group**

Samsung Life Insurance currently holds a 7.4% stake in the South Korean banking companyDGB Financial Group, making it the largest shareholder.[74]

- **Doosan Engine**

Samsung Heavy Industries currently holds a 14.1 percent stake in Doosan Engine, making it the second-biggest shareholder.[75]

- **Korea Aerospace Industries**

Samsung Techwin currently holds a 10 percent stake in Korea Aerospace Industries (KAI). Other major shareholders include the state-owned Korea Finance Corporation (26.75 percent), Hyundai Motor (10 percent) and Doosan (10 percent).[76]

- **MEMC KOREA**

MEMC's joint venture with Samsung Electronics Company, Ltd. In 1990, MEMC entered into a joint venture agreement to construct a silicon plant in Korea.[77]

- **Rambus Incorporated**

Samsung currently owns 4.19 percent of Rambus Incorporated.[78]

- **Renault Samsung Motors**

Samsung currently owns 19.9 percent of the automobile manufacturer Renault Samsung Motors.

- **Seagate Technology**

Samsung currently owns 9.6 percent of Seagate Technology, making it the second-largest shareholder. Under a shareholder agreement, Samsung has the right to nominate an executive to Seagate's Board of Directors.[79]

- **SungJin Geotec**

Samsung Engineering holds a 10 percent stake in Sungjin Geotec, an offshore oil drilling company that is a subsidiary of POSCO.[80]

- **Taylor Energy**

Taylor Energy is an independent American oil company that drills in the Gulf of Mexico based in New Orleans, Louisiana.[81] Samsung Oil & Gas USA Corp., subsidiaries of Samsung, currently owns 20% of Taylor Energy.

Major customers

Major customers of Samsung include:

Royal Dutch Shell

Samsung Heavy Industries will be the sole provider of liquefied natural gas (LNG) storage facilities worth up to US$50 billion to Royal Dutch Shell for the next 15 years.[83][84]

Shell has unveiled plans to build the world's first floating liquefied natural gas (FLNG) platform. At Samsung Heavy Industries' shipyard on Geoje Island in South Korea, work is about to start on a "ship" that, when finished and fully loaded, will weigh 600,000 tonnes – the world's biggest "ship". That is six times as much as the biggest US aircraft carrier.[85]

The world's largest oil and gas project, Sakhalin II- Lunskoye platform under construction. The topside facilities of the LUN-A (Lunskoye) and PA-B (Piltun Astokhskoye) platforms are being built at the Samsung Heavy Industry shipyard in South Korea.[82]

United Arab Emirates government

A consortium of South Korean firms - including Samsung, Korea Electric Power Corp and Hyundai - has won a deal worth 40 billion dollars to build nuclear power plants in the United Arab Emirates.[86]

Ontario government

The government of the Canadian province of Ontario signed off one of the world's largest renewable energy projects, signing a $6.6bn deal that will result in 2,500 MW of new wind and solar energy capacity being built. Under the agreement a consortium – led by Samsung and the Korea Electric Power Corporation – will manage the development of 2,000 MW-worth of new wind farms and 500 MW of solar capacity, while also building a manufacturing supply chain in the province.[87]

Logo

The Samsung Byeolpyo noodles logo, used from late 1938 until replaced in 1958.

The Samsung Group logo, used from late 1969 until replaced in 1979

The Samsung Group logo ("three stars"), used from late 1980 until replaced in 1992

The Samsung Electronics logo, used from late 1980 until replaced in 1992

Samsung's current logo used since 1993.[88]

The current Samsung logo design is intended to emphasize flexibility and simplicity while conveying a dynamic and innovative image through the ellipse, the symbol of the universe and the world stage. The openings on both ends of the ellipse where the letters "S" and "G" are located are intended to illustrate the company's open-mindedness and the desire to communicate with the world. The English rendering is a visual expression of its core corporate vision, excellence in customer service through technology.

The basic color in the logo is blue, the color that the company has had used in its logos for years. The blue color symbolizes stability and reliability, which are precisely what the company wishes to accomplish with its customers. It also stands for social responsibility as a corporate citizen, a company official explained.[89]

Audio logo

Samsung has an audio logo, which consists of the notes E♭, A♭, D♭, E♭. The audio logo was produced by Musikvergnuegen and written by Walter Werzowa.[90][91]

Samsung Medical Center

Samsung donates around US$100 million per annum to the Samsung Medical Center (Korean: □□□□□), a non-profit healthcare provider founded by the group in 1994.[92] Samsung Medical Center incorporates Samsung Seoul Hospital (Korean: □□□□□□), Kangbook Samsung Hospital (Korean: □□□□□□), Samsung Changwon Hospital (Korean: □□□□□□), Samsung Cancer Center (Korean:□□□□□) and Samsung Life Sciences Research Center (Korean: □□□□□□□□□). Samsung Cancer Center, located in Seoul, is the largest cancer center in Asia.[93] Samsung Medical Center and the multinational pharmaceuticals company Pfizer have agreed to collaborate on research to identify the genomic mechanisms responsible for clinical outcomes in hepatocellular carcinoma.[94]

Sponsorships

Olympics

Samsung, which started as a domestic sponsor of the Olympics in Seoul 1988, has been a worldwide Olympic partner since the 1998 Nagano Winter Olympics.[95] Samsung is hoping their role in the London 2012 Olympic Games will provide a "golden moment" for the company's UK reputation, according to Olympic news outlet Around the Rings.[96]

Football

Samsung are the current sponsors of the English Premier League football club Chelsea Football Club. They also sponsor English League One clubs Swindon Town and Leyton Orient.[97]

Price cartels

DRAM price cartel

Samsung was fined EUR 145,728,000 for being part of a price cartel of ten companies for DRAMs which lasted from 1 July 1998 to 15 June 2002. The company received, like most of the other members of the cartel, a 10-% reduction for acknowledging the facts to investigators. Samsung had to pay 90% of their share of the settlement, but Micron avoided payment as a result of having initially revealed the case to investigators.[98]

CRT glass price cartel

Samsung Corning Precision Materials participated in a price cartel of four companies for CRT glass which lasted from 23 February 1999 to 27 December 2004. Samsung Corning Precision Materials received full immunity and was therefore not fined as it revealed the existence of the cartel to the European Commission.[99]

Notes and references

[1] Kelly Olsen (2008-04-22). "Samsung chairman resigns over scandal" (http://web.archive.org/web/20080429210900/http://ap.google.com/article/ALeqM5gmnWKlfgTsbW4n6D9OKnynHdXnhwD906SF9O0). Associated Press via Google News. Archived from the original (http://ap.google.com/article/ALeqM5gmnWKlfgTsbW4n6D9OKnynHdXnhwD906SF9O0) on 2008-04-29. . Retrieved 2008-04-22.

[2] "Samsung Profile 2011" (http://www.samsung.co.kr/samsung/outcome/perfomance.do). Samsung.com. 2010-12-31. . Retrieved 2012-06-10.

[3] http://www.samsung.com/

[4] "Technology - Samsung beats HP to pole position" (http://www.ft.com/cms/s/2/c48d477a-0c3b-11df-8b81-00144feabdc0.html). Ft.com. . Retrieved 2010-09-04.

[5] Economist.com (http://www.economist.com/businessfinance/displayStory.cfm?story_id=13788472) Succession at Samsung – Crowning success

[6] Park, Kyunghee (2009-07-28). "July 29 (Bloomberg) – Samsung Heavy Shares Gain on Shell's Platform Orders (Update1)" (http://www.bloomberg.com/apps/news?pid=newsarchive&sid=aO0FeeTB6_0Y). Bloomberg. . Retrieved 2010-11-11.

[7] "The Top 225 International Contractors2010" (http://enr.construction.com/toplists/InternationalContractors/001-100.asp). Enr.construction.com. 2010-08-25. . Retrieved 2010-11-11.

[8] "Global 500 2009: Industry: - FORTUNE on CNNMoney.com" (http://money.cnn.com/magazines/fortune/global500/2009/industries/183/index.html). Money.cnn.com. 2009-07-20. . Retrieved 2010-09-04.

[9] "The World's Best Amusement Parks" (http://www.forbes.com/2002/03/21/0321feat_6.html). Forbes.com. 2002-03-21. . Retrieved 2010-09-11.

[10] "Cheil Worldwide Inc (030000:Korea SE)" (http://investing.businessweek.com/research/stocks/snapshot/snapshot.asp?ticker=030000:KS). businessweek.com. 2010-09-15. . Retrieved 2010-09-16.

[11] "Agency Family Trees 2010" (http://adage.com/agencyfamilytrees2010/). Advertising Age. 2010-04-26. . Retrieved 2010-09-16.

[12] Hutson, Graham; Richards, Jonathan (17 April 2008). "Samsung chairman charged with tax evasion - Times Online" (http://business.timesonline.co.uk/tol/business/industry_sectors/technology/article3764352.ece). *The Times* (London). . Retrieved 28 February 2011.

[13] "[□ □□□□]<□>□□ □□>□□□□ GDP... □□□ □□□□ –□□□□" (http://www.chosun.com/economy/news/200602/200602130520.html). Chosun.com. . Retrieved 2010-11-11.

[14] "Samsung and its attractions - Asia's new model company" (http://www.economist.com/node/21530984). The Economist. 1 October 2011. . Retrieved 11 January 2012.

[15] "South Korea's economy - What do you do when you reach the top?" (http://www.economist.com/node/21538104). The Economist. 12 November 2011. . Retrieved 11 January 2012.

[16] "□□ 10□ □□ □□□ □□□ □□" (http://www.koreadaily.com/news/read.asp?page=1&branch=NEWS&source=&category=economy.business&art_id=1042338). www.koreadaily.com. 2006-07-10. . Retrieved 2010-09-19.

[17] "History - Corporate Profile - About Samsung - Samsung" (http://www.samsung.com/uk/aboutsamsung/corporateprofile/history06.html). *Samsung Group*. Samsung Group. . Retrieved 13 February 2011.

[18] "Industrial giant's roots tied to nylon products" (http://joongangdaily.joins.com/article/view.asp?aid=2912292). Joongangdaily.joins.com. 2009-11-09. . Retrieved 2011-02-05.

[19] "□□ 40□ □ ..□□ □□ □□□ '□ □'" (http://www.chosun.com/site/data/html_dir/2007/06/19/2007061900326.html). www.chosun.com. 2007-06-19. . Retrieved 2011-02-05.

[20] "SPC-1000" (http://www.old-computers.com/museum/computer.asp?st=1&c=803). old-computers.com. . Retrieved 19 March 2012.

[21] (Korean) Gumisamsung.com (http://www.gumisamsung.com/jsp/gp/GPHistory03.jsp)

[22] "Samsung to celebrate 100th anniversary of late founder" (http://www.koreaherald.com/business/Detail.jsp?newsMLId=20100122000028). koreaherald.com. 2010-03-29. . Retrieved 2011-01-21.

[23] Hansol, Shinsegae Deny Relations with Saehan (http://joongangdaily.joins.com/article/view.asp?aid=1877426) May 24, 2000. Joongangdaily

[24] "Samsung Austin Semiconductor Begins $3.6B Expansion for Advanced Logic Chips" (http://www.austinchamber.com/TheChamber/AboutTheChamber/NewsReleases/2010/SASpressrelease.pdf). Austinchamber.com. 2010-06-09. . Retrieved 2010-09-13.

[25] "Dubai skyscraper symbol of S. Korea's global heights" (http://edition.cnn.com/2009/BUSINESS/10/19/korea.dubai.tower/index.html). CNN. October 19, 2009. . Retrieved 2009-10-19.

[26] Cho, Kevin (2009-04-24). "Samsung Says Hopes of Recovery Are 'Premature' as Profit Falls" (http://www.bloomberg.com/apps/news?pid=20601080&sid=a0wNfW_5OZ5s&refer=asia). Bloomberg. . Retrieved 2010-09-04.

[27] "Samsung buys Sony's entire stake in LCD joint venture" (http://www.bbc.co.uk/news/business-16330877). bbc.co.uk. Dec 26, 2011. .

[28] "Samsung Techwin to spin off camera business" (http://www.reuters.com/article/2008/11/06/samsungtechwin-spinoff-idUSSEO73462008l106). reuters.com. 2008-11-06. . Retrieved 2011-04-05.

[29] "Customers, suppliers & partners" (http://www.rolls-royce.com/korea/en/about/customers_suppliers_partners.jsp). rolls-royce.com. . Retrieved 2011-02-07.

[30] "GEnx-1B Engine Makes its First Flight on Boeing's 787 Dreamliner" (http://www.geae.com/aboutgeae/presscenter/genx/genx_20100616.html). General Electric Company. . Retrieved 2011-02-07.

[31] "South Korea pardons Samsung's ex-chief Lee Kun-hee" (http://news.bbc.co.uk/2/hi/8433297.stm). *BBC News*. 2009-12-29. . Retrieved 2011-09-16.

[32] "South Korea's Pyeongchang Beats Munich, Annecy to Host 2018 Winter Games" (http://www.bloomberg.com/news/2011-07-06/pyeongchang-beats-munich-annecy-to-host-2018-winter-olympics.html). *Bloomberg L.P.*. 2011-07-06. . Retrieved 2011-09-16.

[33] "South Korea's $3.7 Billion Olympic Train to Spur Construction" (http://www.businessweek.com/news/2011-07-20/south-korea-s-3-7-billion-olympic-train-to-spur-construction.html). *Business Week*. 2011-07-20. . Retrieved 2011-09-16.

[34] Business Wire (2011-12-06), *Biogen Idec, Inc. (Massachusetts) (BIIB) Teams With Samsung Corporation on $300 Million Biosimilar Venture* (http://www.biospace.com/news_story.aspx?NewsEntityId=242686) (press release), BioSpace, biospace.com, , retrieved 2012-01-03

[35] Yang, Jun (2011-12-07), "Samsung, Biogen Idec Agree to Set Up $300 Million Venture" (http://www.businessweek.com/news/2011-12-07/samsung-biogen-idec-agree-to-set-up-300-million-venture.html), *Bloomberg Businessweek* (New York City: Bloomberg L.P.): businessweek.com, , retrieved 2012-01-03

[36] "Seagate Completes Acquisition of Samsung's Hard Disk Drive Business" (http://www.seagate.com/ww/v/index.jsp?locale=en-US&name=seagate-completes-aquisition-samsungs-hdd-business-pr&vgnextoid=b201b5c033854310VgnVCM1000001a48090aRCRD). Seagate. . Retrieved 8 February 2012.

[37] "Samsung overtakes Nokia in mobile phone shipments" (http://www.bbc.co.uk/news/business-17865117). BBC News. 27 April 2012. . Retrieved 6 August 2012.

[38] "Samsung overtakes Nokia for Cellphone Lead" (http://www.isuppli.com/Mobile-and-Wireless-Communications/News/Pages/Samsung-Overtakes-Nokia-for-Cellphone-Lead.aspx). . Retrieved 29 April 2012.

[39] "Samsung buys Swedish wireless chip company Nanoradio" (http://www.computerworld.com/s/article/9227658/Samsung_buys_Swedish_wireless_chip_company_Nanoradio). computerworld.com. . Retrieved 11 June 2012.

[40] "Voigtlander & Rollei non-camera items" (http://web.me.com/fwstutterheim/rugarchives/1997-06/00189.html). 1997-06-09. . Retrieved 2011-02-05.

[41] "Basel 96 Watches Take Back the Spotlight" (http://www.jckonline.com/article/284233-Basel_96_Watches_Take_Back_The_Spotlight.php). jckonline.com. 1996-06. . Retrieved 2011-02-05.

[42] "Samsung Loses Attempt to Acquire Fokker" (http://articles.latimes.com/1997-01-01/business/fi-14382_1_maker-fokker). latimes.com. 1997-01-01. . Retrieved 2011-02-05.

[43] "Samsung buys Dutch group in return to M&A" (http://www.ft.com/cms/s/2/65e04374-24b5-11e0-a919-00144feab49a.html#axzz1D1RD7Brp). 1997-06-09. . Retrieved 2011-02-05.

[44] "FUBU Shoes" (http://shoeshowcase.net/fubu-shoes). shoeshowcase.net. . Retrieved 2011-02-05.

[45] "Nomura Wins The Lehman Asia Stakes" (http://www.forbes.com/2008/09/22/nomura-lehman-deal-markets-equity-cx_vk_0922markets03.html). forbes.com. 2008-09-22. . Retrieved 2011-02-07.

[46] "Samsung-Rothschild alliance" (http://www.koreatimes.co.kr/www/news/biz/2010/09/124_33941.html). koreatimes.co.kr. 2008-11-05. . Retrieved 2011-02-07.

[47] Dylan McGrath, EE Times. " Samsung buys MRAM developer Grandis (http://eetimes.com/electronics-news/4218434/Samsung-buys-MRAM-developer-Grandis)." August 2, 2011. Retrieved August 2, 2011.

[48] Chris Preimesberger, eWeek. " Samsung Acquires New-Gen Memory Maker Grandis (http://www.eweek.com/c/a/Data-Storage/Samsung-Acquires-NewGen-Memory-Maker-Grandis-121531/)." August 2, 2011. Retrieved August 2, 2011.

[49] "Samsung to buy Sony half of LCD venture" (http://www.ft.com/intl/cms/s/0/ea650ef6-2fac-11e1-8ad0-00144feabdc0.html). December 26, 2011. .

[50] "Samsung Group plans record $41 billion investment in 2012" (http://www.reuters.com/article/2012/01/17/us-samsung-investment-idUSTRE80G00W20120117). Reuters. 17 January 2012. . Retrieved 8 March 2012.

[51] "□□ 8□□ □□□□ □□□□…□□□ □□ □□□ □□" (http://news.naver.com/main/read.nhn?mode=LSD&mid=sec&sid1=101&oid=003&aid=0003638728#). news.naver.com. 2011-01-13. . Retrieved 2011-08-22.

[52] "□□□□ □□ □□□ □□□ (http://economy.hankooki.com/lpage/industry/201007/e2010073015293447580.htm) □□□ □□□□ □□ □□"]. economy.hankooki.com. 2010-07-30. . Retrieved 2010-09-04.

[53] . hani.co.kr. 2011-04-11. http://www.hani.co.kr/arti/ENGISSUE/74/472384.html. Retrieved 2011-09-18.

[54] The Samsung mission (http://www.samsungmachinetools.com/about/mission/)

[55] "South Korea Starts Grain Venture in Chicago to Secure Supply" (http://www.bloomberg.com/news/2011-04-29/south-korea-starts-grain-venture-in-chicago-to-secure-supply.html). bloomberg.com. . Retrieved 19 March 2012.

[56] "Brooks Automation and Samsung Electronics Announce a Joint Venture" (http://investor.brooks.com/phoenix.zhtml?c=197950&p=irol-newsArticle&ID=840320&highlight=). investor.brooks.com. . Retrieved 28 March 2012.

[57] "POSCO and Subsidiaries" (http://www.rns-pdf.londonstockexchange.com/rns/6677Q_-2009-4-16.pdf). londonstockexchange.com. . Retrieved 06 June 2012.

[58] "POSCO and Subsidiaries" (http://www.rns-pdf.londonstockexchange.com/rns/6677Q_-2009-4-16.pdf). londonstockexchange.com. . Retrieved 06 June 2012.

[59] "Insurance joint venture off to flying start" (http://english.sohu.com/20050707/n226220913.shtml). english.sohu.com. . Retrieved 19 March 2012.

[60] [ht tp://www.livetradingnews.com/biogen-idec-nasdaqbiib-and-samsung-jv-6407.htm "Biogen Idec (NASDAQ:BIIB) and Samsung JV"]. livetradingnews.com. ht tp://www.livetradingnews.com/biogen-idec-nasdaqbiib-and-samsung-jv-6407.htm. Retrieved 19 March 2012.

[61] "Samsung Group, Quintiles Plan $266 Million Venture to Make Biologic Drugs" (http://www.bloomberg.com/news/2011-02-25/samsung-electronics-group-to-form-266-million-biopharmaceuticals-venture.html). bloomberg.com. . Retrieved 19 March 2012.

[62] "Samsung and Sumitomo Chemical to make sapphire substrates for LEDs" (http://www.ledsmagazine.com/news/8/3/24). ledsmagazine.com. . Retrieved 19 March 2012.

[63] "Company Overview of Samsung Thales Co., Ltd." (http://investing.businessweek.com/research/stocks/private/snapshot.asp?privcapId=6464442). businessweek.com. . Retrieved 19 March 2012.

[64] "company/introduce" (http://sdflex.com/company/introduce_1.html). sdflex.com. . Retrieved 19 March 2012.

[65] "Samsung Aerospace, Sermatech Launch Korean JV" (http://english.chosun.com/site/data/html_dir/1999/10/06/1999100661308.html). english.chosun.com. . Retrieved 28 March 2012.

[66] "Siam Samsung Life ready to reawaken" (http://www.bangkokpost.com/business/economics/277211/siam-samsung-life-ready-to-reawaken). bangkokpost.com. . Retrieved 19 March 2012.

[67] "Siltronic-Samsung Joint Venture" (http://www.wacker.com/cms/en/www_archive/www_2008/www_38/38_markets/38_siltronic-samsung/38_siltronic-samsung.jsp). wacker.com. . Retrieved 19 March 2012.

[68] "Completion Ceremony for EDS Production" (http://www.toray.com/aboutus/ourgroup/asia/asi_056.html). toray.com. . Retrieved 28 March 2012.

[69] "Toray/Samsung JV boosts FPD circuit substrate capacity" (http://www.electroiq.com/articles/sst/2008/04/toray-samsung-jv-boosts-fpd-circuit-substrate-capacity.html). electroiq.com. . Retrieved 28 March 2012.

[70] "Alpha's demise thwarts Samsung's processor dreams, analysts say" (http://www.eetimes.com/electronics-news/4102252/Alpha-s-demise-thwarts-Samsung-s-processor-dreams-analysts-say). www.eetimes.com. . Retrieved 2 April 2012.

[71] "[LED EXPO&OLED EXPO 2011 (http://us.aving.net/news/view.php?articleId=197427) An Interview with GE Lighting"]. us.aving.net. . Retrieved 2 April 2012.

[72] "Steel firms in B2B venture" (http://money.cnn.com/2000/05/10/companies/steel_net/). money.cnn.com. . Retrieved 2 April 2012.

[73] "Milestone launch at Brazil's Atlantico Sul" (http://www.marinelog.com/DOCS/NEWSMMIX/2010may00090.html). marinelog.com. . Retrieved 6 March 2012.

[74] "Company Analysis" (http://rdata.youfirst.co.kr/pdf_data/EN_HD_20110201120_00_CC11_21004683_20110201112012_21004849.pdf). rdata.youfirst.co.kr. . Retrieved 19 March 2012.

[75] "Doosan Engine ends 33.2 pct higher on stock market debut" (http://english.yonhapnews.co.kr/business/2011/01/04/57/0503000000AEN20110104008900320F.HTML). yonhapnews.co.kr. . Retrieved 19 March 2012.

[76] "Korea Aerospace sale could prove need for clearer M&A guidelines" (http://www.privateequitykorea.com/ma-news/korea-aerospace-sale-could-prove-need-for-clearer-ma-guidelines). privateequitykorea.com. . Retrieved 19 March 2012.

[77] "MEMC Korea Company" (http://www.memc.com/index.php?view=Chonan). memc.com. . Retrieved 19 March 2012.

[78] "Rambus, Inc. RMBS" (http://cart.morningstar.ca/Quicktakes/owners/MajorShareholders.aspx?t=RMBS®ion=USA&culture=en-US). morningstar.ca. . Retrieved 6 March 2012.

[79] "Seagate to Buy Samsung's Hard-Disk Unit for $1.38 Billion, Build Alliance" (http://www.bloomberg.com/news/2011-04-19/samsung-electronics-seagate-to-combine-computer-hard-disk-drive-business.html). bloomberg.com. . Retrieved 6 March 2012.

[80] "Posco Heavy Industries' may be in the works" (http://121.253.25.110/2012/01/31/2012013105.pdf). Korea JoongAng Daily. . Retrieved 19 March 2012.

[81] "Taylor Energy Sells Gulf of Mexico Assets to Two South Korean Companies" (http://www.reuters.com/article/2008/02/01/idUS180521+01-Feb-2008+BW20080201). reuters.com. . Retrieved 19 March 2012.

[82] "The Russian offshore project "Sakhalin II" is relying on Arma-Chek R" (http://www.armacell.com/www/armacell/INETFAQ.nsf/IDP/330E412A7D4AFD86802570F1003585DC). armacell.com. . Retrieved 2011-02-07.

[83] "Samsung Heavy Industries" (http://www.forbes.com/lists/2009/37/asia-fab-50-09_Samsung-Heavy-Industries_KQZL.html). www.forbes.com. 2009-09-23. . Retrieved 2010-09-13.

[84] "Samsung Heavy Signs Deal with Shell to Build LNG Facilities" (http://www.hellenicshippingnews.com/index.php?option=com_content&task=view&id=58404&Itemid=70). www.hellenicshippingnews.com. 2009-07-31. . Retrieved 2010-09-13.

[85] "The gas platform that will be the world's biggest 'ship'" (http://www.bbc.co.uk/news/science-environment-13709293). bbc.co.uk. . Retrieved 19 March 2012.

[86] "Seoul wins 40-billion-dollar UAE nuclear power deal" (http://www.france24.com/en/20091227-south-korea-wins-40-billion-dollar-united-arab-emirates-nuclear-power-deal). www.france24.com. 2009-12-28. . Retrieved 2010-09-29.

[87] "Korean Companies Anchor Ontario's Green Economy - January 21, 2010" (http://www.premier.gov.on.ca/news/event.php?ItemID=10655). www.premier.gov.on.ca. 2010-01-21. . Retrieved 2010-09-13.

[88] "Case: Samsung 1993" (http://www.corporatebrandmatrix.com/cases.asp?ca_id=55&case=Samsung 1993). .

[89] "□□ 10□ □□ □□□ □□□ □□" (http://www.koreadaily.com/news/read.asp?page=1&branch=NEWS&source=&category=economy.business&art_id=1042338). Koreadaily. . Retrieved 10 July 2006.

[90] "Speaker" (http://audio-branding-academy.org/aba/congress/2k09/speaker/walter-werzowa/). Audio Branding Academy. .

[91] "Logo Video" (http://www.youtube.com/watch?v=KMSnn2C3EAw). Samsung-Youtube. . Retrieved 2011-01-15.

[92] "□□□ □□□□ (http://www.dt.co.kr/contents.html?article_no=2006121302011232655001) □□□□, □□ □□ `□□ □□`... □□□ 1□ "]. www.dt.co.kr. . Retrieved 2010-09-19.

[93] Roberts, Rob (2009-10-26). "AECOM Technology buys Ellerbe Becket" (http://kansascity.bizjournals.com/kansascity/stories/2009/10/26/daily2.html). kansascity.bizjournals.com. . Retrieved 2010-09-19.

[94] "Pfizer And Samsung" (http://www.pfizer.be/pfizer.be/PrintVersion.aspx?Posting={612B1417-EE70-4C51-AD0A-6819653DFBD6}). .

[95] The Olympic Games and Samsung (http://ajw.asahi.com/article/sports/topics/AJ201202170041)

[96] Around the Rings (http://www.aroundtherings.com/articles/view.aspx?id=37194)

[97] Hytner, David (14 May 2009). "Chelsea facing hunt for new shirt sponsor after Samsung stand-off" (http://www.guardian.co.uk/football/2009/may/14/chelsea-samsung-sponsorship-carlo-ancelotti). *The Guardian*. . Retrieved 6 May 2012.

[98] "Antitrust: Commission fines DRAM producers € 331 million for price cartel; reaches first settlement in a cartel case" (http://europa.eu/rapid/pressReleasesAction.do?reference=IP/10/586). *European Commission*. European Commission. . Retrieved 22 April 2012.

[99] "Antitrust: Commission fines producers of CRT glass € 128 million in fourth cartel settlement" (http://europa.eu/rapid/pressReleasesAction.do?reference=IP/11/1214&format=HTML&aged=1&language=EN&guiLanguage=de). European Commission. . Retrieved 22 April 2012.

External links

- Official website (http://www.samsung.com/)
- Samsung Apps (http://www.samsungapps.com/)
- Samsung eStore (https://www.samsungindiaestore.com/Home)
- Samsung (https://twitter.com/samsungtweets) on Twitter

Samsung_Galaxy_Ace_Plus

Samsung Galaxy Ace Plus
GT-S7500

Manufacturer	Samsung Electronics
Series	Galaxy
Compatible networks	**GSM** 850/900/1800/1900 **HSDPA** 7.2 Mbps 900/2100 (Canada **HSPA** 850/1900)[1]
Predecessor	Galaxy Ace
Successor	Samsung Galaxy Ace 2
Type	Smartphone
Form factor	Slate
Dimensions	114.7 x 62.5 x 11.2 mm
Weight	115 gram
Operating system	Android 2.3.6 Gingerbread
CPU	Qualcomm S1 MSM7227A 1 GHz processor
GPU	Adreno 200 GPU
Memory	512 MB
Storage	3 GB
Removable storage	microSD (supports up to 32 GB)
Battery	1,300 mAh, 5.0 Wh, 3.7 V, internal rechargeable Li-ion, user replaceable
Data inputs	Multi-touch, capacitive touchscreen Accelerometer A-GPS Digital compass Proximity sensor Push buttons Capacitive touch-sensitive buttons
Display	TFT LCD, 3.65 in (**unknown operator: u'strong'** mm) diagonal. 320×480 px (158ppi) 16M colors

Rear camera	5.0 megapixel 2592×1944 max, 3x digital zoom, autofocus, LED flash, VGA video recording 640×480 px MPEG4 at 30 fps max.
Compatible media formats	**Audio** MP3, AAC, AAC+, eAAC+ **Video** MP4, H.264, H.263
Ringtones & notifications	Vibration, MP3, WAV
Connectivity	3.5 mm jack Bluetooth v3.0 with A2DP DLNA Stereo FM radio with RDS Micro-USB 2.0 Wi-Fi 802.11 b/g/n
Other	Swype keyboard
SAR	0.84 W/kg

The **Samsung Galaxy Ace Plus** (S7500) is the successor to the mid-range Samsung smartphone Galaxy Ace (S5830). Announced by Samsung on January 3, 2012 for a release in the first quarter of 2012, the Ace Plus will bring a bigger screen and more powerful internals. The phone was released in the UK on March 17, 2012.[2]. The phone is the budget alternative to the Samsung Galaxy Ace 2.

References

[1] "Telus Samsung Galaxy Ace Details" (http://www.telusmobility.com/en/ON/samsung_galaxy_ace/index.shtml). telusmobility.com. . Retrieved 2011-11-17.

[2] http://www.technobuffalo.com/companies/samsung/samsung-galaxy-ace-plus-arrives-in-u-k-tomorrow-march-17/

Samsung_Galaxy_S_Advance

Samsung I9070 Galaxy S Advance

Manufacturer	Samsung Electronics
Series	Galaxy
Compatible networks	2G: GSM 850 / 900 / 1800 / 1900 3G: HSDPA 850 / 900 / 1900 / 2100
First released	2012, April
Introductory price	~ $325[1]
Availability by country	Asia, Middle-East, Europe, USA
Predecessor	Samsung Galaxy S, Samsung Galaxy SL I9003
Dimensions	123.2 x 63 x 9.69 mm
Weight	120 g
Operating system	Android OS v2.3.6
CPU	1GHZ Dual-core NovaThor U8500 chipset
GPU	ARM Mali-400MP (single-core)
Memory	768MB RAM
Storage	8/16GB built memory
Removable storage	microSD up to 32GB
Battery	1500 mAh
Display	Super AMOLED, 4 inches, 800x480 pixels resolution
Rear camera	5MP + HD 720p video recording
Front camera	Yes (1.3MP)
Compatible media formats	MP3, AAC, AAC+, WMA, WAV, MP4, H.264, 3Gp, DivX/Xvid support
Ringtones & notifications	WAV/MP3

Samsung GT-I9070 Galaxy S Advance is an Android smartphone which was announced in January 2012 and released in April 2012. The Galaxy S Advance tech specifications includes a Super AMOLED Display with the size of 4 inches, dual-core processor with 1 GHz clock frequency and a rear camera with resolution of 5 MP and a front camera for video calls.

reference(s)

[1] http://www.amazon.com/Samsung-I9070-Galaxy-Advance-Quad-band/dp/B008277H40/ref=sr_1_2?s=electronics&ie=UTF8&qid=1341524785&sr=1-2&keywords=Galaxy+S+Advance

Sources

- *Samsung Galaxy S Advance Review* (http://reviews.cnet.co.uk/mobile-phones/samsung-galaxy-s-advance-review-50007145/) at Cnet
- *Samsung Galaxy S Advance tech specs* (http://www.gsmarena.com/samsung_i9070_galaxy_s_advance-4469.php) at GSMArena.com
- *Samsung Galaxy S Advance Review* (http://www.phonearena.com/reviews/Samsung-Galaxy-S-Advance-Review_id3043) at phonoarena.com

Article Sources and Contributors

Samsung_Galaxy_Ace_2 *Source*: http://en.wikipedia.org/w/index.php?title=Samsung_Galaxy_Ace_2 *Contributors*: Chienlit, Cysioland, Graeme Bartlett, Gtwfan52, J36miles, Sun Creator, Trivialist, Zqzqqzqzqzqqzq, 19 anonymous edits

Smartphone *Source*: http://en.wikipedia.org/w/index.php?title=Smartphone *Contributors*: 16@r, 3Coins, A Man In Black, A Quest For Knowledge, A123a10, A8UDI, AKA MBG, ALSLUG, AVM, Aaditya 7, Acalamari, Acather96, Aclassifier, Acolin f, Adib5271, Adriatikus, Aeons, Ahoerstemeier, Aillema, Ajsandboch, Akuyume, Alansohn, Alex890, Alf Boggis, Alf.laylah.wa.laylah, AlistairMcMillan, Allen3, Amarendra.Avinash, AndrewSpec, Andries, Andros 1337, Angela, Angstygangsta, AniRaptor2001, Ansible, Anupam, Arab Dynasty, Arafael, Argentina subcampeon 1789, Arny, Arp120, Arunsingh16, Atama, Atenyi, Atul.ecn, Auntof6, Aurelie Branchereau-Giles, Avoided, BD2412, Bahaltener, Bakshi41c, Balazer, Baronnet, Beland, Bendes, Bevo74, Bflaherty77, Bhny, Bibijee, Biker Biker, BillHaywood, BlkStarr, Blobby54, Bomazi, Bovineone, Branddobbe, Brianreading, Brittanybee1121, Broccoli, CLW, CRJO-CRJO, Cactusframe, Cadiomals, Caj27, Calabe1992, Calcwiki, CalumCook234, CanadianPenguin, Canalteen, Catan wen, CatherineMunro, Cellcom, ChaChaFut, Changshui88, Charivari, Chris the speller, Ciphers, Citizensmith, Cleared as filed, Closedmouth, Clovis Sangrail, Cmf, Cmlewan, Cmp101, Codetiger, Comaleaf, CommonsDelinker, Commontimect, Cool Hand Luke, Coolaaron88, CoolingGibbon, Cornea503, Cst17, Currab, Cvdr, Cxtom, Cyruslei, DVdm, DaisyField, Dale Arnett, Dan6hell66, Danhash, Daniel.finnan, DaveChild, Davidbengtenglund, Dcljr, Dcxf, Ddavid2005, Deane@gooroos.com, Delphii, DerBorg, Deviceapps, Diannaa, Diego Moya, Discospinster, DivineAlpha, Dmarquard, Dreambringer, Drobatch, E.au, EVula, EWikist, Editor182, Editrrr, Elassint, Electronicbazaarnz, Electronicguru1, Emma23 K, Epolk, Eraserhead1, Erc, Erianna, Eshade, Esrever, EugeneZelenko, Eugenia loli, Evice, Evilruletheworld96, Evolutiondb, F1MotoGPWRC, FCYTravis, Farolif, Father Goose, Fcassia, Feinoha, Feydey, Fhontoy, FinnsDeal, Flowanda, Fourdee, Fraggle81, Furrybeagle, Fuzheado, GTBacchus, Galadh, Garant^ ^, GateKeeper, Geologyguy, Georgy90, Gfannick, Giftlite, Giggett, Gilliam, Ginsengbomb, Giraffedata, Glacialfox, Glennwells, Gmumru, Gogo Dodo, GoingBatty, Gold Hat, Gomm, Gookey, Gordon Ecker, GraemeL, Grafen, Grahamperrin, Grajales, Graywolf, Graywolfmoon1, Gregory Heffley, Gregorysalt, Greswik, Groink, Gsarwa, Guy Harris, Gwernol, HDCase, Ha us 70, Haakon, Hadiceberg, Halloween.mac, Hamiltha, Hammersoft, Hardylane, Harmil, Harryzilber, Haseo9999, Hede2000, Henrik, HereToHelp, Hervegirod, Hobartimus, Hu12, Human.v2.0, Huntersquid, HuskyMoon, Huw Powell, HybridBoy, Hydrogen Iodide, Hydrox, Hypnotist uk, I already forgot, I, Podius, Ianb, Ianboudreault, Ianereed, Icedwater, Ignis Fatuus, Illegal Operation, Illyria05, Im Buff, Immunmotbluescreen, Imroy, Imsilly123, In2thats12, Indefatigable, Interframe, InternetMeme, Intershark, IntrigueBlue, Ipsign, Irky, Ishdarian, ItsZippy, J, J.delanoy, JCDenton2052, JCrue, JNW, Jaesi25, Jagdterrier, Jakupju, JamesWeb, JamieS93, Jantangring, Jaspermaz, Jeffq, Jekyllhide, Jerome Charles Potts, Jerryobject, Jfdwolff, Jim.henderson, Jimmy Bergmark, Jobin RV, JoeSmack, John of Reading, JohnChrysostom, JonHarder, Jonabbey, Jondel, JonoP, Joseph Solis in Australia, Josh Parris, Jostake, Jsherwood0, Jsribeiro, Juan M. Gonzalez, Jusdafax, Just another editor, Justin Mauger, Jwojdylo, Jæs, Kajowi, Kariteh, KarmaGeddon, Katieh5584, Kbdank71, Kcomstock, Kellen`, KelleyCook, Kentynet, Kevin Dorner, Kingpin13, Kinst, Kkm010, Klokbaske, KnowBuddy, Koavf, Kozuch, Kraftlos, Kresp0, Kudpung, KungFuMonkey, Kuru, Kwaichi, LaVieEntiere, Lai888, Lambtron, Larkhill97, Laurasmith70308, Lazulilasher, LeilaniLad, Lenin1991, Lester, Leszek Jańczuk, Leujohn, Level plus, Lewispb, Lifes g00d 561, Lilac Soul, Limequat, Little Mountain 5, Little Professor, Logan, Loser997, Lotje, LrdChaos, Ltomuta, Lukan4ica, Lun Esex, Lun4tic, M0rphzone, MER-C, MMuzammils, Mac, Mac John Concord, Magnus.de, Maine12329, Malik Shabazz, Mange01, Maniacgeorge, Manop, Manway, Maproom, Marcisjustice, Marco.difresco, Marcus Qwertyus, Mark Kim, Mark91, Marksza, Marrowmonkey, Martarius, Masgatotkaca, Materialscientist, Mathiastck, Mato, Maury Markowitz, Maxí, Mcduck, Mckote, Mdrejhon, Mdwh, Meehawl, Meelar, Memaster3, Mephistophelian, MetaManFromTomorrow, Miaow Miaow, Mikael Häggström, Mike Dillon, Mike Linksvayer, Mikehelms, Mild Bill Hiccup, MillenniumB, Minghong, Miquonranger03, Mix Bouda-Lycaon, Mlg07e, Moberg, ModWilliam, Monaarora84, Mononomic, Morphh, Mortense, Mosmof, Mr. Strong Bad, MrOllie, Mvjs, My76Strat, MySchizoBuddy, MyronAub, Myscrnnm, Nakakapagpabagabag, Nathan94124, NathanYDR, Navacell, Nazrich, Ne0Freedom, Nealmcb, Neo Ogilvy, Neoarchon, Neoguy999115, Nerdeff, Netrapt, Nick Number, Nickolai 420, Night Gyr, Nightscream, Nitro.ajb, Nixdorf, Nneonneo, Nopetro, Nurg, Obey, Obiwankenobi, Ohnoitsjamie, Oknazevad, OlavN, Old port, Oli Filth, Omardarwish10, Omegatron, Oontay, OpenToppedBus, P.Marlow, PS., PTSE, Parintachin, Patrick, Pats1, Patvac-chs, Paul 1953, Pbrown111, Pdahomepage, Petershank, Peyre, Phatalflaw, Phatom87, Phearson, Philip Trueman, Piano non troppo, Pingveno, Piotrek54321, Pluma, Pmarshal, Pmlineditor, Pol098, Pol430, Poor Yorick, Posix memalign, Prathameshsasane, Prillen, Pro66, Pvanheus, Pyfan, Quebec99, Querencia, R'n'B, RA0808, Radiier, Radon210, Rafael.sp, Ramu50, RedWolf, Reliablesoft, Repetition, Res2216firestar, ReverseEngineered, Reyk, Rgreed, Rhobite, Riadlem, Rich Farmbrough, Richardguk, Richiekim, Riki, Rollins83, Ron2, Ronz, RoyBoy, Ruchir257, Rursus, SF007, SPQRobin, Sadegh87, Sainath468, Samad120, Sameer singh17, Samwb123, Sandstein, Sapibobo, Sayid98, Schrödinger's Cake, ScottyBerg, Scrtcwlvl, Sctechlaw, Sdfisher, Sdrazfar, Sean13zz, Sebastian Mandrean, Sebindcruz, Secretsorry, Serg3d2, Sfacets, Sfan00 IMG, Sfsmartp, Shalroth, Shawnc, Sheehan, Signalhead, Siliconov, Singerdg1, SirJibby, Sjcramer, Skatebiker, Skintigh, Skittle, Slitchfield, Smart1954, Smartphonespakistan, Smmgeek, Snigbrook, Soliloquial, SpaniardGR, Spellmaster, Starofwonder, Stephen B Streater, Stephen Turner, Stjson, SuperHamster, Surenkarapetyan, Suriel1981, Suwatest, Swoof, TVS99, Tahir mq, Taka76, Takamaxa, Taketa, Tangent1000, Taskinen, TastyPoutine, Tatterfly, Tcomotcom, Technopat, Tesi1700, Tfgbd, ThaWhistle, ThaddeusB, The Pink Oboe, The Thing That Should Not Be, The Wild Falcon, The-apathy, Theaveng, Thestick, Thiled, Thisma, Thumperward, Ticell, Tide rolls, Tikru8, TimSE, Tinton5, Tklaer, Tmuller2, Tobias Bergemann, Togopogo, Tokyogamer, Tomas.turek18, Tombomp, Tommy2010, Tommytentimes, Tony1, Toussaint, Tpbradbury, Trasz, Treekids, Tuju, UCLATre, UnrealG, Urod, Uzume, Vegaswikian, Verkinto, Verne Equinox, Versus22, Vkem, Vrenator, Waidawut, Wallace1966, Waqas Hussain, Wbm1058, Westie4321, WhisperToMe, Whkoh, WiZZLa, Wideangle, Wiki admi, Wikimhb, WikipedianYknOK, Wikipelli, Winged-stone, Wireless Buddy, Wknight94, Woohookitty, XGrape, Xajel, Xezbeth, Xrobertcmx, Yean3d, Youxiarock, Yug, Yunshui, Zach Vega, Zanter, ZimZalaBim, ZipoB, Zpetro, [] [] [] , 1200 anonymous edits

Android_(operating_system) *Source*: http://en.wikipedia.org/w/index.php?title=Android_%28operating_system%29 *Contributors*: 336, 777sms, 9th jinchuriki, A bit iffy, A.sutton, A520, A5b, A665321, Aadadurov, Abhishek191288, Abhishekitmbm, Abledsoe78, Abrahami, AbstractClass, Aceleo, Acery, Adambiswanger1, Adamjacobd, Adamwatters, Adi19956, Adileader, AdjustablePliers, Adm.Wiggin, Afriza, Aftekology, Aftershave, Agentlame, Akbarzpro, Akshayjain123, AladdinSE, Alboran, Alejo2083, Alex, AlexKucherenko, AlexMS, Alexey Izbyshev, Alexius08, Alfredo ougaowen, Alfstar1997, Ali'i, Aliendude5300, Alisha.4m, AlistairMcMillan, Allen Moore, Allstarecho, Alunphillips, Alvestrand, Amarenderjannu, Amatulic, Ambictus, Amckern, Ameliorate!, Ancheta Wis, Andareed, Anderssl, Andreas Bischoff, Andreas Carter, Andrejavus, Andresfi, Andrew Delong, AndrewHowse, Andrewkantor, Android4.0, Androidliscence, Androidmids, Andyjsmith, Anindya Bakshi, Ankitasdeveloper, AnonMoos, Anoopan, Anoopmichael, Anthonynon, Antnee, Aoeuser, Apobilgin, Aquarat, Aradius, Aranci, Arc Orion, Arcanis, Archangelsk, ArgetlahmSource, Arghya139, Arichnad, Arjun G. Menon, Arthas01, Aryamanjain, Aryndar, Ash Crow, Ashishjain999, Ashwin18, Astonmartini, Asymmetric, Athzai Khaine, Atlantia, AtteL, Attilios, Audi152, Audriusa, Automate, AwamerT, Axl, Ayd00, Ayd000, Ayd86, Aydceri, Aydcery00, Aydcery86, Aydchery00, Aydin00, BY.Apps, Bagelfat101010, Bahua, Bangbang.S, Banksbr2, Barek, Baronnet, Barte, Bayonetblaha, Bbaumer, Bbisgard, Bdesham, Bedna, Beland, Ben Tillman, Bender235, Benjaminb, Benlisquare, Berelv, Betmenko, Bevo, Bhny, Bilbo571, Binarybits, Bios Element, Blackfireshocker, Blahbabe61, Blethering Scot, Blindwaves, Blogfactor, Blowdart, Bmwtroll, Bohr999, Bomazi, Bonadea, Bongwarrior, Bosqueschool, Bovineone, Bpave777, Brandorr, Brianreading, Brianski, Briantist, Bryan.burgers, BucsWeb, Bungalowbill, Bweono, C628, CCalo, CJLL Wright, CJMiller, CPGustafson511, CRGreathouse, CTF83!, Cacophony, Caltas, Can't sleep, clown will eat me, CaribDigita, Carlton.northern, Carrlos, CaseyBorders, CastAStone, Causa sui, Cbmaster, Cbr1000f, Cburnett, Ch Th Jo, Chainz, Chancer1001, ChaosData, Chardot, Charles.h.white@gmail.com, CharlesC, Cheekeong123, Cherie327, Cherkash, Chezi-Schlaff, Chiles Malesters, Chillpenguin, Chillum, Chirags, Chr1syr, Chris Bainbridge, Chris the speller, Chris.p.wu, ChrisHeller, Chrismiceli, Chrispilot2293, Christian75, Ciphergoth, Ciphers, ClaudeReigns, Clovis Sangrail, Coffee, Colonies Chris, CommonsDelinker, ComputerGeek706, Conan, Conti, Conzorz, Coolbho3000, Coolstoryhansel, CoordinateFreak, Corevette, Cornlad, Costa Discordia, Count Chockula, Cpl Syx, CrabbyPatrick, Crackerspeanut12, Craigbarnes85, Craigbrass, Cresdajv, Crysb, Csrempert, Cybercobra, Cyrotux, D-Notice, D20sheets, DGMDGM, DHN, DStoykov, DVdm, Daabomb, Daev, Dale Arnett, Dancter, Dandyandroid, Danger, Daniel.Cardenas, DanielPharos, DarTar, Dark-Fire, Darkspy945, DarrenW, Darrenm540, David Edgar, David Woodward, Davidbengtenglund, Davidhorman, Dawnseeker2000, Dbachmann, Dcxf, Deattitude, Degorr, Demysc, Denniss, Dennisthe2, Desbest, Dgw, Dhaluza, Diamondland, Diannaa, Diblidabliduu, Diego Moya, Diego.viola, Digana, Digilee, Dingar, Dismas, Dj.cowan, Dlrohrer2003, Dmit, DmitryKo, Docu, Dontmitchell, Douglaswth, Download, Dpupkov, Dr.Soft, Dra.vladvamp, DragonLord, Drbreznjev, Dreaded Walrus, Drogonov, Drrll, Dsh13, Dsrivallabha, DudeThinking, Dudyk, Dueynz, Dvyjones, Dziedrius, E258, E2eamon, EEMIV, ESkog, Eaefremov, Eagle-slayor, Eapache, Ebe123, Echeese, Eciepecie, Ed Burnette, Ed Poor, EdBever, Edoe, Edward, Ej159, Ekerazha, Electron9, Elektron, Eleman, Elronxenu, Emurphy42, EngineerFromVega, Enigmaticland, Ennustaja, EonOmega, Erc, Eric 324, ErkinBatu, Erwin Mulialim, Esebi95, Espertus, Essayemyoung4009, Estemi, Ettrig, EugeneKay, Eugrus, Exien, Explorer25, Faddykeyboard, Fanatic.manav, Fangfufu, Fastily, FatalError, Fattmann, Fbtjock, Feedmecereal, Ferengi, Feudonym, Ffinder, Filmore, Finalius, Firefoxian, Fish and karate, Flatterworld, Fleminra, FlieGerFaUstMe262, Flintb, Flohack, Fluffylouis, Found5dollar, Fox hyx, Fran McCrory, Frankie, Frap, FredTubale, Freddicus, Free French, Friginator, Frood, Fryn, Fsamuels, Furrykef, Fxhomie, Gabriel A. Zorrilla, Gainesk, Gaius Cornelius, Galaxytab, Gallagher783, Gary King, Gautamkishore, Gboxdance, Gchangetok, Gdm, Geary, Gegorg, George Ponderevo, Gerhman, Gh5046, Ghepeu, Ghettoblaster, Ghost650, Glany222, GlasGhost, GlassCobra, Glen 3B, Goa103, Gogo Dodo, Gogoloid, GoingBatty, Gokberks, GoldKanga, Golftheman, Good Olfactory, Googlemobileplatform, Googlesubculture, Gordon Ecker, GorillaWarfare, Gouranga Gupta, GraemeL, Grafen, Graft, Graig123, Grandscribe, GreenpeaceUbuntuMan, Gregconquest, Gregory Heffley, Grika, Gronky, Grshiplett, Gsarwa, Gscshoyru, Gsonnenf, Gu1dry, Gudeldar, Gugu2903, Guitarguy99081, Gurch, Guru4321, Guyjohnston, Guzzyron, Gyro Copter, H4lfN3ls0n, HJ Mitchell, Haakon, Hacheema, Haggisfarm, Hammersoft, Hanifbbz, Hannes1983, HardCorwen, Harimohan07, Harizotoh9, Harp, Haseo9999, Hcaandersen, Headinthedoor, Hedge777, Henriok, Henry W. Schmitt, Herakleitoszefesu, Hervegirod, Hgb asicwizard, Hockeyc, Hominid, Hoss789, HotXRock, Hotcrocodile, Howlingmadhowie, Htchien, Htinlinn90, Hu12, Hucz, Hughcharlesparker, Huku-chan, Hutchinsonam, Hydrox, Hymek, I Feel Tired, I, Podius, I5bala, IBoy2G, IGEL, IRWolfie-, ISTB351, Ian1337, IceHunter, Icep, Icydesign, Iggymwangi, Ijon, Iknowyourider, IlPisano, Illegal Operation, Imagine Reason, Immunmotbluescreen, Imroy, Indianstar, InternetMeme, Invenio, InverseHypercube, InvertedPendulum, Ionistii, Iridescent, Irishguy, Irislia, Ironmagma, Isaacwaller, Island Monkey, Iuhkjhk87y678, Ivant, Ivario, J.delanoy, JAMJAM1666, JCDenton2052, JEN9841, JHunterJ, JLMCGE01, Jack007, Jacob Poon, JacobSheehy, Jacobmathias, Jadden14, Jairodz, Jaizovic, Jalabi99, Jamadagni, James Foster, JamesBWatson, JamesNBarnes, Jamgraham88, Jamougha, Jarble, JaredMT, Jasper Deng, Javachan, JavierCane, JavierMC, Jb0807, Jboyens, Jbreckenridge, Jcogbil, Jdthood, JeR, Jeff G., Jeffq, Jeffrey Sharkey, Jeffwang, Jenova20, Jerebin, JeremyA, Jeromeds99, Jerrinsg, Jerryobject, Jesant13, Jessica23, Jeysaba, Jhonnyx1000, Jiess, Jim1138, Jimmin, Jimthing, Jimv1983, Jinmyo, Jmcdon10, Jmecimore, JobiWan144, Joconnor, Joeblow398847232, Joemalt1832, Johantheghost, John Ericson, JohnSawyer, Johnconorryan, Johndburger, JohnnotJon, Jokonek, Jonabbey, Jonathan-Morris711, JonathonSimister, Jonkerz, Jontintinjordan, Jopo, Jorge Stolfi, Joriki, JosJuice, Josh.e.stroud, JoshDuffMan, Jpvinall, Jreferee, Jrishel, Jtangsw, Jtfcobra, Jubeidono, Julesd, Jurisnipper, Jusses2, Jvosika, Jwkilgore, KAMiKAZOW, KDesk, KSEltar, Kaicarver, Kaisersushi, Karam.Anthony.K, Karthickmad, Katherine, Katoh, Kawasakik, KayoWikiP069, Kenny Strawn, Kentyman, Kepwick, Kevin James Field, Kevthegreat55, Kforeman1, Khalid hassani, Khanayub1986, KhrOn0s, Kiddington, Kien64, KimDabelsteinPetersen, Kingdowney, Kingpin13, Kinkate18nic, Kiore, Kitsunegami, Kkm010, Klemen Kocjancic, Klingoncowboy4, Kmdowns, Knud Winckelmann, Koavf, Kokken Tor, Koman90, Komitsuki, Kontar, Korkut00, Korkut000, Kozuch, Kraftlos, Krazywrath, Krc1185, Kris cs1, Kronox android, Ksyrie, KumardipSarkar, Kungming2, Kushal one, Kylelnny,

Kylesamani, LSUniverse, LafinJack, LancerEvolution :, Lanilsson, Larrymcp, LarsHolmberg, LarsPensjo, Lbstone27, Legoboy920, Lenar, Lesmin, Lester, Lethe, Leuko, Levineps, Lexischemen, Lfcohen, Libcub, LightSpeed3, Lightenoughtotravel, Lightmouse, Lindamilton, Lindberg, Lkt1126, Llancast, Llewelyn MT, Logan, Logical Cowboy, Lokpest, Longhornkate, LookingGlass, Lopifalko, LorenzoB, Lotje, Lovetinkle, Lucas.Yamanishi, Luckerr, Lukini, Lun Esex, M0sia1, MER-C, MKar, MZMcBride, Mac, Macungie, Madeincat, Magioladitis, Mahanga, Male1979, Manop, Mantrik00, Mappum, Marc Lacoste, Marcus Brute, Marcus Qwertyus, Marcus2020, Mardus, Marek69, Mark Renier, Mark0528, Markmcwiggins, Marko Gargenta, Markpb91, Marksbark, Marqueed, Martin.komunide.com, Mastrsushi, Materialscientist, Mathewsherdil, Matt Darby, Matthew0028, MatthewBurton, MattieTK, Mattkap, Mattkap2, Maulikdave05, Maurice Carbonaro, Mauripop, Maxdeutc, Maximus06, Maxviwe, Maxí, McGeddon, Mcld, Mdikici, Meepzip, Meersmaj, Melab-1, Melizg, Melmann, Mendaliv, Mentifisto, Mephistophelian, Messiisking, MetaManFromTomorrow, Mewtu, Mharen, Michaelplourde66, Midgetman433, Mikael Häggström, Mike Rosoft, Mike.lifeguard, Milan Keršláger, Mild Bill Hiccup, Millstream3, Miltonhowe, Mimihitam, Mindmatrix, Minterior, Mirabilos, Miserlou, Mistral Mktg, Mistral Solutions, Mkouklis, Mobilecushion, Mobilepush, Modamoda, Mogism, Mohanpram, Mohit kesarwani, Moneytoo, Moocha, Mordka, MoreNet, MoreThings, Morian27, Morning277, Mortense, Mr. Met 13, MrGALL, MrOllie, Ms2ger, Muelaner, Mugsywwiii, Mugunth Kumar, Mutchy126, MyNameWasTaken, Myas012, Myscrnnm, N2e, N5iln, NTox, NYKevin, Naddy, Nagy Dániel, Nagytam, Nahado, Namures, Nantasatria, Nathanloop, Ne0Freedom, NeMeSiS, Nealmcb, NeilN, Neinsun, NetHunter, Newone, Newsoxy, NexuSix, Nexus26, Nick Garvey, Nico357, Nicolas Love, Nigelj, Nightscream, NiklasBr, Nikpapag, Ninly, Njonji, Nodekeeper, Nogburt, Noloader, Noozgroop, Norm mit, Now wiki, Nuclearmoose, Nuujinn, Nyco, Obiwankenobi, Ofennell, Ohaaron, Ohconfucius, Ohnoitsjamie, OlavN, Old Number7, Oldmokmok, Oleg Alexandrov, Oli Filth, Omiqa, Omshivaprakash, Orange Suede Sofa, Originalwana, OsamaK, OspreyPL, OwenBlacker, Oxwil, P.Shack, P2jones, PILZI, Papatenor, Pascal.Honore, Patrick, Paulmlieberman, Paulscrawl, Pdfpdf, Pelago, Pelthais, Peter712, Peterkagey, Pgan002, Phalinshah, Phatom87, Philip Trueman, PhosphoricX, Phy1729, Piast93, PieterDeBruijn, Pigr8, Pinball22, Pinecar, Pjedicke, Pkkasu, Plankhead, PlantRunner, Plarem, Plop, Pluma, Pmod, Pmyteh, Pokstad, Pol098, PolarYukon, Pomegranate, Pooooooooo123, Potentials, Pr4733k, PriceChild, Prius 2, Privateboz, Procedure, ProfPolySci45, Profvladthethird, Prolog, Prosfilaes, Prototypecreative, Pryanni, Psantora, Pseudomonas, PutzfetzenORG, Pvanderlee, Pwnage97, Quarkgluonsoup, Quartermaster, Quebec99, Queenmomcat, Quoth, Qwyrxian, RScheiber, Raburton, Rachel263, Raghualluri, Rahil.kassamali, Rajanbalana, Rajeshsweb, Ral725, Ralfsmouse, RameshaLB, Ramonrabello, Random name, Randomname66, Rapjul, Rapomon, Ravipokemon, Raysonho, Rborghese, Rchandra, RcketScientist, Reaper Eternal, RedHillian, Redekopmark, Reebsauce, Reedy, RegenerateThis, Renergade1, RenniePet, Res2216firestar, Resplendent, Rich Farmbrough, Richard Arthur Norton (1958-), Richi, Richiekim, Riffic, Rigelt, Riki, Ringbang, RingtailedFox, Rjwilmsi, Rmanke, Robbrown, Robert Moyse, RobertMfromLI, Roberth Edberg, Roberto.larcher, Robferrer, Robzz, Rocboronat, RockMFR, Rockysmile11, Rod92p, Rodeosmurf, Roguegeek, Roif456, Ronnies1312, RotaryAce, RoyBoy, Royce, Rprpr, Rsrikanth05, Rugops, Runtime, Rush2009, Rwalker, Ryan8374, RyanQuinlan, Ryry17354, RzR, Rzęsor, S1lencing, SF007, Sachinchavan.in, Sagarwal1981, Saggy84, Saifuddinap, Sailsbystars, Sainath468, Salamurai, Salazasu, Salvio giuliano, SamJohnston, Samdman95, Samkass, Samoscool, Samuh, Sandstein, SarekOfVulcan, Sasank, Sayden, Sbmeirow, Scampy11, Sciencewatcher, Scientus, Scl98029, ScottyWZ, Seanjacksontc, Searchmaven, Secretlondon, SephirothXIIIX, Seven.cardwell, SeyedKevin, Sfm 7, Shachar, Shadez08, Shadowjams, ShakataGaNai, Shaolinx, SharkD, Shaswat Narendra, Shevett, SidP, Sierra1bravo, Sigma2488, Sijil cv, Silvio Marano, Simonrleung, Simple Bob, Sirlancer, Six words, SixSix, Sjames1, Sjl0523, Skier Dude, Skierpage, Skudo900630, Skype565, Slakr, Slatedorg, Sleepy Sentry, Sligocki, Small.is.powerful, Smashville, Smitty, Smurfy, Smyth, Snakeskincowboy, SoWhy, Socialmaven1, Solinym, Solipsys, Solomon Douglas, Soma6, Some jerk on the Internet, Someguy1221, Southpitt, Sp33dyphil, Spaghetti64, Speculatrix, Spiel, Sreyan, Sriram sh, Staka, StaticGull, Steel, Stefan, Stephenb, Stephenwanjau, Steve03Mills, Steve1428, Stevedel7, Steveklein, Steven Walling, Stevenbz9, Stevenmitchell, Stevenwagner, StewieK, Strcat, Subbu, Suckystraw, Suction Man, SudoGhost, Sukael, Sunnypsyop, Sunray, Superm401, Suzals3, SvGeloven, Svetovid, Svick, Swampyank, Swatjester, Sygmoral, Sylvainchevalierfu, Syndicate, Syp, T-Nod, T-Rex84, TMO KOTOR, TXI59, Tagrb03f, Tahir mq, Tahitiville, Taras, TarzanJr, TastyCakes, Tavilis, Tazio99, Tbhotch, Tbird20d, Tcncv, Tedder, Tedickey, Teeks99, Teginc, Teleprinter Sleuth, Teles, Tgeairn, The Anome, The Letter J, The.rahul.nair, The359, TheEditrix2, TheTechFan, TheWishy, Thealexweb, Theanphibian, Theartfullodger, Thecurran91, Themfromspace, Thesamami, Thingg, ThomasWilson2, Thorwald, Thumperward, Thunder Wolf, Thunderbird8, Thüringer, Tide rolls, Timeshifter, TimmmmCam, Timneu22, TobiasPersson, Tobziez, Tomchen1989, Tomlzz1, Tondi5, Tony Sidaway, TonyW, TorQue Astur, Torqueing, Traal, Tracer9999, TrbleClef, Trebek Skates, Trefork, Tri400, Troed, Trollered32145, Trollskini, Trusilver, Tsriopensourceblueprints, Tuxcantfly, Tweisbach, Txaggiemichael, TyA, U5K0, UKER, UU, Uirauna, Ujimatcha, Ulric1313, Unamed102, Unknownwarrior33, Urashimataro, Urfavoritemija, User931, Usmanahmed25, Vadmium, Vahid83, Vald, Van helsing, Varlaam, Vcelloho, Venona, Victorpardosi, Vincenzo.romano, Voidvector, VoluntarySlave, Vrenator, Vujke, WakiMiko, Walter Görlitz, Walterlmitchell3, Waltonkbbl, Watchcars, Wbison3, Wednesday Next, Wello95, Werbej, Werdna, Wertydm, Wesleyarchbell, WhatMeWork, Whatiknow, Wickedjacob, Wifuk, WikiLaurent, Wikigod, Wikipedian2009, Wild mine, William Leadford, Williameis, Windofkeltia, Wintermute115, Wizardist, Wknight94, Wlindley, Woohookitty, Woolfy123, Writermonique, Wtmitchell, Wwoods, XJamRastafire, Xavierorr, Xcrivener, XdaLive, Xhienne, Xiutwel-0003, Xnamkcor, Xomm, Xrobau, Xsspider, Xx3nvyxx, Y2kcrazyjoker4, YICbaby, Yadavjpr, Yahia.barie, Yamla, Yaohong3914, YasharF, Yellowdesk, Yiosie2356, Yngvarr, Yousou, YuMaNuMa, Yug, Yunshui, Yuriybrisk, ZacBowling, Zaratoustra, Zbutler7, Zeldex, Zero sharp, ZimZalaBim, Zipcodeman, ZirconiumTwice, Zorak950, Zouzzou, Zundark, Zunmun, ZzRayzZ, ^musaz, Ævar Arnfjörð Bjarmason, Λ, Սահակ, . □□□□ . □□□□□□□□ , 2516 anonymous edits

Samsung *Source*: http://en.wikipedia.org/w/index.php?title=Samsung *Contributors*: 10metreh, 16@r, 28bytes, 5 albert square, 660gd4qo, 7596cakpn, 9-11 suicide bomber, A8UDI, Achowat, Acroterion, Adam Zivner, AdamFouracre, Adrian90, Afluegel, Agentbla, Agusm266, Ahoerstemeier, Aido2002, AirFrance358, Aitias, Akbg, Alansohn, AlexanderKaras, Ali, Alphabet55, Alphachimp, American1234, AnOddName, Andonic, Andrea105, AndresTM, AnimatedZebra, Anirvan, Antonio Lopez, Anzk, Apeman2001, Appleby, Arafel6, Ardiedan1995, ArgleBargleSmargle, Arthena, Aruhnka, Arvindn, Asdjfsdljsdj, Aspects, Avi.seth8, Avono, Azoz839, BD2412, Bakuryuu, Ballerjg69, Ballsinjaws, Balrogi, Banej, Barbreuer, Bason0, Battulafamily, Bboymonkeyking, Bbx, Beardo, Begoon, Belmonter, Bender235, Bentendo24, Bep-and-matt, Bidgee, Bidsell, Bilbo571, Billythekilly, Bjbushpig, Bjk28, Blahblah5555, Blhilbrands, Boccobrock, Bongomatic, BorgHunter, Bovineone, Bqrius, Brandmeister, Brandon47, Brazuca100, Brett Hwang, Brewbg, Brittany Ka, Brrunkle, Byeonggwan, Calcwatch, CallMeHenry, Calltech, CambridgeBayWeather, Cameron Scott, Candyshpownr, Canopus1, Capricorn42, Carioca, Carmichael, Caspian blue, Ceyockey, Chaitanya.lala, Charles Gaudette, Charlesdrakew, Chazburrows, Chconline, Chdbtm, Check two you, Chloride, Chris the speller, Chuunen Baka, Ckatz, ClariT, ClementSeveillac, Cloudcounts, Cocojaco, CommonsDelinker, ConMan, Conay, Coolhandscot, Core0002, Courcelles, Crap777, Crispmuncher, CronusXT, Cs-wolves, Culnacreann, Cyber 217, Cydevil38, D, D 2 da a, D-Ball10, D3sisted, DAJF, DARTH SIDIOUS 2, DMacks, DVD R W, DVdm, DaMan92, Daaavid, Dac04, Daeyung, Damienchock, Dancter, DandanxD, Daniel C, Daniel Olsen, Darkpoet, Darkwind, DarlingApertures, DaronDierkes, Daroosterhawk, Darth Mike, Daumdaumdaum, Davecrosby uk, David Biddulph, David Levy, David44357, DavidFarmbrough, Dbal124, DeadEyeArrow, Deiaemeth, Deiz, Dekisugi, Delldot, Denisarona, Dennis Bratland, DepartedUser4, Desagwan, Destron Commander, Dev1n, Dfrg.msc, Diannaa, Dispenser, Djr xi, Do it, Doksuri, Dolafootlong, Domino666, Dommar, Donald12, Doniago, Doulos Christos, Download, Drummer868, Dual Freq, Dwa652, Dwim79, DylanGMills, E.aisthebest, E2eamon, Easternknight, Ec17s05, Edward, Eliz81, Elockid, Elvey3:16, EncMstr, Engas, Epbr123, Erik9, Escottf, Esrever, Eugenebata, Excirial, Excretion, Exeunt, Ext.ppa, Eyeoffire, FV alternate, FastLizard4, FayssalF, Feneeth of Borg, Fetx2002, Fieldday-sunday, Filmboyz, Filmluv, Firsfron, FishHeadAbcd, Flewis, Flyguy649, Fnlayson, Folksong, Fourohfour, Frenchman113, Frickative, Frietjes, Frysun, Gaborgasko, Gadgetkorea.com, Gadiez, Gary King, Geeoharee, Geniac, Georgeofdelhi, Gezuniga, Ghfjdksla123, Gilliam, Ginsengbomb, Giraffedata, Glane23, Gmoore96, Gmosoti, Godsmodo, Goffrie, Gogo Dodo, GoingBatty, Goldstar012, Grace624chun, Grant Lee, Grayshi, Greenshed, GregorB, Gsarwa, Gunmetal Angel, Gus Polly, H magic, Ha98574, HarrisonB, Heartnseoul, Hede2000, Henry Park, Heroeswithmetaphors, Hetar, Historiographer, Hogibear, Hooterslover, Hopiakuta, Hroðulf, Hu12, Hudwu, Hughnewman7, Humorahead01, Husond, Huss34huss34, Hvn0413, Iamnununu, Icarus 198, Idolescent, Igob8a, Ilha Youn, Ilovesit06, Imnotplaying, InCypher, Indizio, Ingolfson, Invincible Ninja, Iohannes Animosus, Iridescent, Irishguy, Ironman5247, Irstu, Isaac Crumm, Izehar, J.delanoy, JBsupreme, JForget, JGXenite, JIP, JNW, Jab843, Jack fa el do, Jackfork, Jacques de Vincent, Jadeyybbe, Jahnda, Jamcib, JamesAM, JamieBalfour07, JamieS93, Jandalhandler, Janus657, Janviermichelle, Japandamonium, Jeff G., Jeff3000, Jeffc96, Jeffq, Jehochman, Jennycheil, Jerem43, Jeremyb, Jim Douglas, Jimmy Slade, Jj137, Jklamo, John of Reading, Johnuniq, Jon C., Jonathanjouty, Jons63, Joseph Solis in Australia, Jovianeye, Jpark1088, Juicebox6980, Justme89, Jweiss11, K92700259, KAMiKAZOW, KEIM, KUsam, Kappa, Kcandrsn, KennethHan, Kevintown, Kgf0, KimJooYeon, Kingj123, KingsTCJB, Kintetsubuffalo, Kiranantovadassery, Kiteinthewind, Kkm010, Klushka, Kookyunii, Krballer, Krish Dulal, Kristoff Vernard 2788, Krtek2125, Kuebie, Kukja33, Kye999, Kymagnus, L Kensington, LFCTrav, LG4761, Lakshmix, Larry laptop, LeCire, LeaveSleaves, Lemento, LittleWink, Liyang, Ljubisaaa27, Lkxcnldsf, Logan, Lorial235, Lotje, LtPowers, Luccas, M-le-mot-dit, MDfoo, MER-C, MG42COD, MMuzammils, MONGO, Mabdul, Maddie26, Malc82, Manngnth, Marcopolis, Mark83, Mark91, Markisgreen, Martarius, Masterofwin, MatchStickEleven, Materialscientist, Mathiastck, Matthew Stannard, Matto123456, Mayamore, Mazingspidrman, Mdann52, Medjaw1, MelbourneStar, Mgc0wiki, Miami33139, Michael Hardy, Mieciu K, Mighty librarian, Mightyoat, Mikaly, Mikegatopoulos, Mild Bill Hiccup, MilesFrmOrdnary, MilkMiruku, Milkmooney, Miremare, Mirmo!, Miruru, Missayou88, Mksoccercity, Monkeyman, Moondyne, Moonriddengirl, MorrisS, Mr. Stradivarius, MrOllie, MrShamrock, Mrarchitectkim, Mrmanwhochangesstuff, Multivariable, Mumyeong, Musclehamster199, Mushu150, Mycroft7, Mysdaao, Mystique 309, Myu0246, NTox, Nakagawa0, Nasnema, NawlinWiki, Neil Clancy, NeilN, Nepenthes, Nickcerda, Nikai, Nima1024, Noaster2, Nobhead1000, Noirthekiller, NoseNuggets, Nottora2, Noveltyghost, Nthep, Nutellaficker1, OhanaUnited, Ohnoitsjamie, OkakiMCMLXX, Old Guard, Oli Filth, Omaer1016, Oneiros, Ottawakismet, OverlordQ, Owlpamy, Oyo321, PC78, PTSE, Pak21, Parthrana, Pascal.Tesson, Patchfinder, Patelj603, PattonWarKing, Paulinho28, Paulleeis, Paulo3099, Paulsabater, Pds0101, Peripitus, Persian Poet Gal, Petiatil, Pfalcone, Pgdn963, Pharaoh of the Wizards, Phearson, Philip Trueman, Piano non troppo, Pichu1988, Piercemacca, PigFlu Oink, Pikiwyn, Plattler01, Polk540, Potatoefun!, Poweron, Ppntori, Pro translator, Prolog, Q43, Qasamaan, QuiteUnusual, Qxz, R'n'B, RFerreira, RHaworth, Raamin, Raime, Raiseplus, Ralphbk, Rangoon11, Razorflame, Rcawsey, Reformsamsung, Reliableforever, Reliancepowercoin, Requestion, Res2216firestar, Rich Farmbrough, Richiekim, Ricochet 100, Rimmington01, Rjanag, Rjwilmsi, Rkm12587, Roakr, Rob1974, Rockysmile11, Romanapizza, Romatoto, RoseTech, Ruinswikipages, Runforit97, RxS, Ryan Norton, S, S3000, SCEhardt, SKS2K6, Sado, Safoo, Salih, Sam John0, SamJohnston, Samsara, SamsungBR, Samtex9, Sapphic, SarahKD, SarahStierch, Schzmo, Scrodimcboogerballs, Scxnwa, Seaphoto, Seotraf, Seventynights, Sf5xeplus, Shaunvonronn, Shawn in Montreal, Shawnc, Shortride, Sirbobcharles, Sitk, Skids 3580, Sknab, Slavon37, Slo-mo, Smint83, Smsarmad, Snowolfd4, Soccerboystuart, Softjuice, Solbob, Sonhye, Sonofthebeach, South Bay, Spearhead666, Species8473, Speedboy Salesman, SpeedyGonsales, Squibman, StarScream1007, StaticGull, Steenies, Stephen, Stevage, Stevenmitchell, Stifle, Styrofoam1994, Sundaryourfriend, Superk1a, Superp, Svick, Swat0120, TDogg310, TROLLOLL master, Tagus, Tankiona, TechGeek70, Tedats, Tehw1k1, Tekblurb, Tempodivalse, Terano30, Terence, Tharsaile, The undertow, The wub, TheLastOne, Thehelpfulone, Thehuntall, Therealdantheman, Thingg, Tholme, Thompsontough, Tide rolls, TigerShark, Tikiwont, Timllmixit, Tmeski, Tnaniua, Tndrckr, Toasted Onion, Toddblarg, Tonman777, Toreau, Toshio Yamaguchi, Towel401, Tracer9999, Traf22, Trakesht, Tranquilantus, Treisijs, Trevj, Truism77, Trusilver, Truthbedarned, Tsung, Tufta, TutterMouse, TweaknewsNathan, Ty1993, Umask777, Uncle Milty, UnitedStatesian, Unsellt, Vald, Varundbest10, Vegaswikian, Victorysda, Visviva, Vivek.shetty, Vobgro, Vrenator, W Tanoto, W1k1m0derat0r564, WCAWiki, WJetChao, WOSlinker, Walkingbeef, Wandering-teacher, War wizard90, We hope, Webdinger, WelshMatt, Wesley Moy, Western Pines, Wibbble, WikHead, WikiDan61, WikiLeon, WikiPuppies, Wikikath, Winbuyer, Winston365, Wondergirls, Woohookitty, Wordsmith, Worldorg, Woysy, Wtg87, Wwandsumww, X9TORX, XRobby, Xosé, Yamla, Yaninass2, Yellowdesk, Ykhwong, Ymir 3749, Yomamma4, Yongjik, Yorkshiresky, Youngjoon Shin, Yowuza, Yoyoitsneuman22, Yuyudevil, ZBAUGH, Zachlipton, Zhou Yu, Zidonuke, Ziggymaster, ZimZalaBim, Ziziziii, Zoakrwkd, Zzuuzz, §, שבויר, □□□ . □□□ , 1728 anonymous edits

Samsung_Galaxy_Ace_Plus *Source*: http://en.wikipedia.org/w/index.php?title=Samsung_Galaxy_Ace_Plus *Contributors*: Amged.hasan, Bearcat, Biasoli, CommonsDelinker, Feudonym, Jeikei, Katharineamy, Luiswtc73, Mtelee, NatGertler, PamD, Skype565, Theboywonder, Trivialist, WPSamson, Wikien2009, Woohookitty, Zeromusx, 14 anonymous edits

Samsung_Galaxy_S_Advance *Source*: http://en.wikipedia.org/w/index.php?title=Samsung_Galaxy_S_Advance *Contributors*: Edoderoo, Faramarz, Rwalker, Trivialist, 4 anonymous edits

Image Sources, Licenses and Contributors

File:IBM Simon Personal Communicator.png *Source*: http://en.wikipedia.org/w/index.php?title=File:IBM_Simon_Personal_Communicator.png *License*: Public Domain *Contributors*: Tcomotcom, 3 anonymous edits

File:Nokia 9210.jpg *Source*: http://en.wikipedia.org/w/index.php?title=File:Nokia_9210.jpg *License*: GNU Free Documentation License *Contributors*: J-P Kärnä

File:Htc Touch Pro2 Georgy.JPG *Source*: http://en.wikipedia.org/w/index.php?title=File:Htc_Touch_Pro2_Georgy.JPG *License*: Creative Commons Attribution-Sharealike 3.0,2.5,2.0,1.0 *Contributors*: Georgy90

File:Original iPhone docked.jpg *Source*: http://en.wikipedia.org/w/index.php?title=File:Original_iPhone_docked.jpg *License*: Creative Commons Attribution-Sharealike 2.0 *Contributors*: Andrew from London, UK

File:Samsung Galaxy S III Pebble Blue Wikipedia.png *Source*: http://en.wikipedia.org/w/index.php?title=File:Samsung_Galaxy_S_III_Pebble_Blue_Wikipedia.png *License*: Creative Commons Attribution-Sharealike 3.0 *Contributors*: User:Illythr

File:Global Mobile Applications Store Revenue.svg *Source*: http://en.wikipedia.org/w/index.php?title=File:Global_Mobile_Applications_Store_Revenue.svg *License*: Creative Commons Attribution-Sharealike 3.0 *Contributors*: MySchizoBuddy

File:World Wide Smartphone Sales Share.png *Source*: http://en.wikipedia.org/w/index.php?title=File:World_Wide_Smartphone_Sales_Share.png *License*: GNU Free Documentation License *Contributors*: Smartmo

File:Android robot.svg *Source*: http://en.wikipedia.org/w/index.php?title=File:Android_robot.svg *License*: Creative Commons Attribution 3.0 *Contributors*: Google

File:Android logo.png *Source*: http://en.wikipedia.org/w/index.php?title=File:Android_logo.png *License*: Public Domain *Contributors*: Google

File:Android 4.1 on the Galaxy Nexus.jpeg *Source*: http://en.wikipedia.org/w/index.php?title=File:Android_4.1_on_the_Galaxy_Nexus.jpeg *License*: unknown *Contributors*: Android Open Source project

File:G1, Nexus One, Nexus S, Galaxy Nexus.jpg *Source*: http://en.wikipedia.org/w/index.php?title=File:G1,_Nexus_One,_Nexus_S,_Galaxy_Nexus.jpg *License*: Creative Commons Attribution-Sharealike 2.0 *Contributors*: FlickreviewR, Infrogmation, Materialscientist, OspreyPL, 2 anonymous edits

File:Android-System-Architecture.svg *Source*: http://en.wikipedia.org/w/index.php?title=File:Android-System-Architecture.svg *License*: Creative Commons Attribution-Sharealike 3.0 *Contributors*: User:Smieh

File:Android home.png *Source*: http://en.wikipedia.org/w/index.php?title=File:Android_home.png *License*: GNU General Public License *Contributors*: Unamed102

File:PlayStorePermissions.png *Source*: http://en.wikipedia.org/w/index.php?title=File:PlayStorePermissions.png *License*: unknown *Contributors*: Smurfy

File:Android chart.png *Source*: http://en.wikipedia.org/w/index.php?title=File:Android_chart.png *License*: Creative Commons Attribution 2.5 *Contributors*: Android Open Source project

File:Samsung Logo.svg *Source*: http://en.wikipedia.org/w/index.php?title=File:Samsung_Logo.svg *License*: Public Domain *Contributors*: Samsung

Image:□ □ □ □ .jpg *Source*: http://en.wikipedia.org/w/index.php?title=File:□ □ □ □ .jpg *License*: Public Domain *Contributors*: Mumyeong

Image:SPC-1000.JPG *Source*: http://en.wikipedia.org/w/index.php?title=File:SPC-1000.JPG *License*: Public Domain *Contributors*: (User:Zanny)

Image:Pan-samsung2-error corrections.png *Source*: http://en.wikipedia.org/w/index.php?title=File:Pan-samsung2-error_corrections.png *License*: Public Domain *Contributors*: Zoakrwkd

File:Expo 2012 Samsung pavilion.JPG *Source*: http://en.wikipedia.org/w/index.php?title=File:Expo_2012_Samsung_pavilion.JPG *License*: Creative Commons Attribution-Sharealike 3.0 *Contributors*: hyolee2

Image:Ssbld002.jpg *Source*: http://en.wikipedia.org/w/index.php?title=File:Ssbld002.jpg *License*: Public Domain *Contributors*: Original uploader was Marcopolis at en.wikipedia

File:Korea HQ Office.jpg *Source*: http://en.wikipedia.org/w/index.php?title=File:Korea_HQ_Office.jpg *License*: Creative Commons Attribution 3.0 *Contributors*: Secl

Image:Lunskoye.jpeg *Source*: http://en.wikipedia.org/w/index.php?title=File:Lunskoye.jpeg *License*: Public Domain *Contributors*: Ssr, Towel401

File:Past(1938) samsung logo.PNG *Source*: http://en.wikipedia.org/w/index.php?title=File:Past(1938)_samsung_logo.PNG *License*: Trademarked *Contributors*: samsung

File:Past(1969-79) samsung logo.PNG *Source*: http://en.wikipedia.org/w/index.php?title=File:Past(1969-79)_samsung_logo.PNG *License*: Trademarked *Contributors*: samsung

File:Past samsung.PNG *Source*: http://en.wikipedia.org/w/index.php?title=File:Past_samsung.PNG *License*: Trademarked *Contributors*: samsung

File:Samsung-old.gif *Source*: http://en.wikipedia.org/w/index.php?title=File:Samsung-old.gif *License*: Trademarked *Contributors*: unknown

GNU Free Documentation License Version 1.2, November 2002 Copyright (C) 2000,2001,2002 Free Software Foundation, Inc. 59 Temple Place, Suite 330, Boston, MA 02111-1307 USA Everyone is permitted to copy and distribute verbatim copies of this license document, but changing it is not allowed.

0. PREAMBLE

The purpose of this License is to make a manual, textbook, or other functional and useful document "free" in the sense of freedom: to assure everyone the effective freedom to copy and redistribute it, with or without modifying it, either commercially or noncommercially. Secondarily, this License preserves for the author and publisher a way to get credit for their work, while not being considered responsible for modifications made by others. This License is a kind of "copyleft", which means that derivative works of the document must themselves be free in the same sense. It complements the GNU General Public License, which is a copyleft license designed for free software. We have designed this License in order to use it for manuals for free software, because free software needs free documentation: a free program should come with manuals providing the same freedoms that the software does. But this License is not limited to software manuals; it can be used for any textual work, regardless of subject matter or whether it is published as a printed book. We recommend this License principally for works whose purpose is instruction or reference.

1. APPLICABILITY AND DEFINITIONS

This License applies to any manual or other work, in any medium, that contains a notice placed by the copyright holder saying it can be distributed under the terms of this License. Such a notice grants a world-wide, royalty-free license, unlimited in duration, to use that work under the conditions stated herein. The "Document", below, refers to any such manual or work. Any member of the public is a licensee, and is addressed as "you". You accept the license if you copy, modify or distribute the work in a way requiring permission under copyright law. A "Modified Version" of the Document means any work containing the Document or a portion of it, either copied verbatim, or with modifications and/or translated into another language. A "Secondary Section" is a named appendix or a front-matter section of the Document that deals exclusively with the relationship of the publishers or authors of the Document to the Document's overall subject (or to related matters) and contains nothing that could fall directly within that overall subject. (Thus, if the Document is in part a textbook of mathematics, a Secondary Section may not explain any mathematics.) The relationship could be a matter of historical connection with the subject or with related matters, or of legal, commercial, philosophical, ethical or political position regarding them. The "Invariant Sections" are certain Secondary Sections whose titles are designated, as being those of Invariant Sections, in the notice that says that the Document is released under this License. If a section does not fit the above definition of Secondary then it is not allowed to be designated as Invariant. The Document may contain zero Invariant Sections. If the Document does not identify any Invariant Sections then there are none. The "Cover Texts" are certain short passages of text that are listed, as Front-Cover Texts or Back-Cover Texts, in the notice that says that the Document is released under this License. A Front-Cover Text may be at most 5 words, and a Back-Cover Text may be at most 25 words. A "Transparent" copy of the Document means a machine-readable copy, represented in a format whose specification is available to the general public, that is suitable for revising the document straightforwardly with generic text editors or (for images composed of pixels) generic paint programs or (for drawings) some widely available drawing editor, and that is suitable for input to text formatters or for automatic translation to a variety of formats suitable for input to text formatters. A copy made in an otherwise Transparent file format whose markup, or absence of markup, has been arranged to thwart or discourage subsequent modification by readers is not Transparent. An image format is not Transparent if used for any substantial amount of text. A copy that is not "Transparent" is called "Opaque". Examples of suitable formats for Transparent copies include plain ASCII without markup, Texinfo input format, LaTeX input format, SGML or XML using a publicly available DTD, and standard-conforming simple HTML, PostScript or PDF designed for human modification. Examples of transparent image formats include PNG, XCF and JPG. Opaque formats include proprietary formats that can be read and edited only by proprietary word processors, SGML or XML for which the DTD and/or processing tools are not generally available, and the machine-generated HTML, PostScript or PDF produced by some word processors for output purposes only. The "Title Page" means, for a printed book, the title page itself, plus such following pages as are needed to hold, legibly, the material this License requires to appear in the title page. For works in formats which do not have any title page as such, "Title Page" means the text near the most prominent appearance of the work's title, preceding the beginning of the body of the text. A section "Entitled XYZ" means a named subunit of the Document whose title either is precisely XYZ or contains XYZ in parentheses following text that translates XYZ in another language. (Here XYZ stands for a specific section name mentioned below, such as "Acknowledgements", "Dedications", "Endorsements", or "History".) To "Preserve the Title" of such a section when you modify the Document means that it remains a section "Entitled XYZ" according to this definition. The Document may include Warranty Disclaimers next to the notice which states that this License applies to the Document. These Warranty Disclaimers are considered to be included by reference in this License, but only as regards disclaiming warranties: any other implication that these Warranty Disclaimers may have is void and has no effect on the meaning of this License.

2. VERBATIM COPYING

You may copy and distribute the Document in any medium, either commercially or noncommercially, provided that this License, the copyright notices, and the license notice saying this License applies to the Document are reproduced in all copies, and that you add no other conditions whatsoever to those of this License. You may not use technical measures to obstruct or control the reading or further copying of the copies you make or distribute. However, you may accept compensation in exchange for copies. If you distribute a large enough number of copies you must also follow the conditions in section 3. You may also lend copies, under the same conditions stated above, and you may publicly display copies.

3. COPYING IN QUANTITY

If you publish printed copies (or copies in media that commonly have printed covers) of the Document, numbering more than 100, and the Document's license notice requires Cover Texts, you must enclose the copies in covers that carry, clearly and legibly, all these Cover Texts: Front-Cover Texts on the front cover, and Back-Cover Texts on the back cover. Both covers must also clearly and legibly identify you as the publisher of these copies. The front cover must present the full title with all words of the title equally prominent and visible. You may add other material on the covers in addition. Copying with changes limited to the covers, as long as they preserve the title of the Document and satisfy these conditions, can be treated as verbatim copying in other respects. If the required texts for either cover are too voluminous to fit legibly, you should put the first ones listed (as many as fit reasonably) on the actual cover, and continue the rest onto adjacent pages. If you publish or distribute Opaque copies of the Document numbering more than 100, you must either include a machine-readable Transparent copy along with each Opaque copy, or state in or with each Opaque copy a computer-network location from which the general network-using public has access to download using public-standard network protocols a complete Transparent copy of the Document, free of added material. If you use the latter option, you must take reasonably prudent steps, when you begin distribution of Opaque copies in quantity, to ensure that this Transparent copy will remain thus accessible at the stated location until at least one year after the last time you distribute an Opaque copy (directly or through your agents or retailers) of that edition to the public. It is requested, but not required, that you contact the authors of the Document well before redistributing any large number of copies, to give them a chance to provide you with an updated version of the Document.

4. MODIFICATIONS

You may copy and distribute a Modified Version of the Document under the conditions of sections 2 and 3 above, provided that you release the Modified Version under precisely this License, with the Modified Version filling the role of the Document, thus licensing distribution and modification of the Modified Version to whoever possesses a copy of it. In addition, you must do these things in the Modified Version: A. Use in the Title Page (and on the covers, if any) a title distinct from that of the Document, and from those of previous versions (which should, if there were any, be listed in the History section of the Document). You may use the same title as a previous version if the original publisher of that version gives permission. B. List on the Title Page, as authors, one or more persons or entities responsible for authorship of the modifications in the Modified Version, together with at least five of the principal authors of the Document (all of its principal authors, if it has fewer than five), unless they release you from this requirement. C. State on the Title page the name of the publisher of the Modified Version, as the publisher. D. Preserve all the copyright notices of the Document. E. Add an appropriate copyright notice for your modifications adjacent to the other copyright notices. F. Include, immediately after the copyright notices, a license notice giving the public permission to use the Modified Version under the terms of this License, in the form shown in the Addendum below. G. Preserve in that license notice the full lists of Invariant Sections and required Cover Texts given in the Document's license notice. H. Include an unaltered copy of this License. I. Preserve the section Entitled "History", Preserve its Title, and add to it an item stating at least the title, year, new authors, and publisher of the Modified Version as given on the Title Page. If there is no section Entitled "History" in the Document, create one stating the title, year, authors, and publisher of the Document as given on its Title Page, then add an item describing the Modified Version as stated in the previous sentence. J. Preserve the network location, if any, given in the Document for public access to a Transparent copy of the Document, and likewise the network locations given in the Document for previous versions it was based on. These may be placed in the "History" section. You may omit a network location for a work that was published at least four years before the Document itself, or if the original publisher of the version it refers to gives permission. K. For any section Entitled "Acknowledgements" or "Dedications", Preserve the Title of the section, and preserve in the section all the substance and tone of each of the contributor acknowledgements and/or dedications given therein. L. Preserve all the Invariant Sections of the Document, unaltered in their text and in their titles. Section numbers or the equivalent are not considered part of the section titles. M. Delete any section Entitled "Endorsements". Such a section may not be included in the Modified Version. N. Do not retitle any existing section to be Entitled "Endorsements" or to conflict in title with any Invariant Section. O. Preserve any Warranty Disclaimers. If the Modified Version includes new front-matter sections or appendices that qualify as Secondary Sections and contain no material copied from the Document, you may at your option designate some or all of these sections as invariant. To do this, add their titles to the list of Invariant Sections in the Modified Version's license notice. These titles must be distinct from any other section titles. You may add a section Entitled "Endorsements", provided it contains nothing but endorsements of your Modified Version by various parties--for example, statements of peer review or that the text has been approved by an organization as the authoritative definition of a standard. You may add a passage of up to five words as a Front-Cover Text, and a passage of up to 25 words as a Back-Cover Text, to the end of the list of Cover Texts in the Modified Version. Only one passage of Front-Cover Text and one of Back-Cover Text may be added by (or through arrangements made by) any one entity. If the Document already includes a cover text for the same cover, previously added by you or by arrangement made by the same entity you are acting on behalf of, you may not add another; but you may replace the old one, on explicit permission from the previous publisher that added the old one. The author(s) and publisher(s) of the Document do not by this License give permission to use their names for publicity for or to assert or imply endorsement of any Modified Version.

5. COMBINING DOCUMENTS

You may combine the Document with other documents released under this License, under the terms defined in section 4 above for modified versions, provided that you include in the combination all of the Invariant Sections of all of the original documents, unmodified, and list them all as Invariant Sections of your combined work in its license notice, and that you preserve all their Warranty Disclaimers. The combined work need only contain one copy of this License, and multiple identical Invariant Sections may be replaced with a single copy. If there are multiple Invariant Sections with the same name but different contents, make the title of each such section unique by adding at the end of it, in parentheses, the name of the original author or publisher of that section if known, or else a unique number. Make the same adjustment to the section titles in the list of Invariant Sections in the license notice of the combined work. In the combination, you must combine any sections Entitled "History" in the various original documents, forming one section Entitled "History"; likewise combine any sections Entitled "Acknowledgements", and any sections Entitled "Dedications". You must delete all sections Entitled "Endorsements".

6. COLLECTIONS OF DOCUMENTS

You may make a collection consisting of the Document and other documents released under this License, and replace the individual copies of this License in the various documents with a single copy that is included in the collection, provided that you follow the rules of this License for verbatim copying of each of the documents in all other respects. You may extract a single document from such a collection, and distribute it individually under this License, provided you insert a copy of this License into the extracted document, and follow this License in all other respects regarding verbatim copying of that document.

7. AGGREGATION WITH INDEPENDENT WORKS

A compilation of the Document or its derivatives with other separate and independent documents or works, in or on a volume of a storage or distribution medium, is called an "aggregate" if the copyright resulting from the compilation is not used to limit the legal rights of the compilation's users beyond what the individual works permit. When the Document is included in an aggregate, this License does not apply to the other works in the aggregate which are not themselves derivative works of the Document. If the Cover Text requirement of section 3 is applicable to these copies of the Document, then if the Document is less than one half of the entire aggregate, the Document's Cover Texts may be placed on covers that bracket the Document within the aggregate, or the electronic equivalent of covers if the Document is in electronic form. Otherwise they must appear on printed covers that bracket the whole aggregate.

8. TRANSLATION

Translation is considered a kind of modification, so you may distribute translations of the Document under the terms of section 4. Replacing Invariant Sections with translations requires special permission from their copyright holders, but you may include translations of some or all Invariant Sections in addition to the original versions of these Invariant Sections. You may include a translation of this License, and all the license notices in the Document, and any Warranty Disclaimers, provided that you also include the original English version of this License and the original versions of those notices and disclaimers. In case of a disagreement between the translation and the original version of this License or a notice or disclaimer, the original version will prevail. If a section in the Document is Entitled "Acknowledgements", "Dedications", or "History", the requirement (section 4) to Preserve its Title (section 1) will typically require changing the actual title.

9. TERMINATION

You may not copy, modify, sublicense, or distribute the Document except as expressly provided for under this License. Any other attempt to copy, modify, sublicense or distribute the Document is void, and will automatically terminate your rights under this License. However, parties who have received copies, or rights, from you under this License will not have their licenses terminated so long as such parties remain in full compliance.

10. FUTURE REVISIONS OF THIS LICENSE

The Free Software Foundation may publish new, revised versions of the GNU Free Documentation License from time to time. Such new versions will be similar in spirit to the present version, but may differ in detail to address new problems or concerns. See http://www.gnu.org/copyleft/. Each version of the License is given a distinguishing version number. If the Document specifies that a particular numbered version of this License "or any later version" applies to it, you have the option of following the terms and conditions either of that specified version or of any later version that has been published (not as a draft) by the Free Software Foundation. If the Document does not specify a version number of this License, you may choose any version ever published (not as a draft) by the Free Software Foundation. ADDENDUM: How to use this License for your documents To use this License in a document you have written, include a copy of the License in the document and put the following copyright and license notices just after the title page: Copyright (c) YEAR YOUR NAME. Permission is granted to copy, distribute and/or modify this document under the terms of the GNU Free Documentation License, Version 1.2 or any later version published by the Free Software Foundation; with no Invariant Sections, no Front-Cover Texts, and no Back-Cover Texts. A copy of the license is included in the section entitled "GNU Free Documentation License". If you have Invariant Sections, Front-Cover Texts and Back-Cover Texts, replace the "with...Texts." line with this: with the Invariant Sections being LIST THEIR TITLES, with the Front-Cover Texts being LIST, and with the Back-Cover Texts being LIST. If you have Invariant Sections without Cover Texts, or some other combination of the three, merge those two alternatives to suit the situation. If your document contains nontrivial examples of program code, we recommend releasing these examples in parallel under your choice of free software license, such as the GNU General Public License, to permit their use in free software.